# Preparing Scheduled Desludging

# Preparing Scheduled Desludging

Produced by USAID – IUWASH PLUS Program

Rudy Yuwono, Foort Bustraan, Budi Darmawan,
Endro Adinugroho, Nugroho Andwiwinarno and Ika Fransiska

IWA PUBLISHING

**Published by** **IWA Publishing**
**Unit 104–105, Export Building**
**1 Clove Crescent**
**London E14 2BA, UK**
Telephone: +44 (0)20 7654 5500
Fax: +44 (0)20 7654 5555
Email: publications@iwap.co.uk
Web: www.iwapublishing.com

First published 2021


*British Library Cataloguing in Publication Data*
A CIP catalogue record for this book is available from the British Library

ISBN: 9781789062274 (paperback)
ISBN: 9781789062281 (eBook)

This eBook was made Open Access in May 2021

# Contents


**Foreword** ... ix

**Preface** ... xi

**Introduction** ... xiii

***Understand scheduled desludging*** ... *1*

0.1 A Mandatory Desludging ... 1
0.2 Desludge and Transport ... 3
0.3 The Many Benefits ... 4
0.4 Seven Aspects to Consider ... 5
0.5 Thirteen Development Steps ... 7

**Step 1**
***Identify local strengths*** ... *9*

1.1 Analyze the Gap ... 9
1.2 Rely on Available Data ... 9
1.3 'Go' or 'No Go' ... 11

**Step 2**
***Agree on the basics*** ... *13*

2.1 Establish Common Perception ... 13
2.2 Define Outcomes ... 14
2.3 Set Area and Time ... 14
2.4 Propose Institutions ... 16
2.5 Recommend Financial Principle ... 16

**Step 3**
***Make initial concept*** ........ ***17***

3.1 Just an Illustration ........ 17
3.2 Estimate Operations ........ 18
3.3 Calculate Basic Cost ........ 19
3.4 Strive for Profit ........ 22

**Step 4**
***Get the green light*** ........ ***25***

4.1 Permission to Continue ........ 25
4.2 Present Persuasively ........ 25
4.3 Show it Graphically ........ 27
4.4 Record the Decision ........ 29

**Step 5**
***Assess targets*** ........ ***31***

5.1 Use Current Information ........ 31
5.2 Perform Survey ........ 32
5.3 Collect Data ........ 33
5.4 Analyze for Information ........ 34
5.5 Use for Other Purposes ........ 37

**Step 6**
***Design operations*** ........ ***39***

6.1 Classify Targets ........ 39
6.2 Divide Service Zones ........ 40
6.3 Confirm Desludging Period ........ 40
6.4 Specify Level of Service ........ 42
6.5 Determine Modes of Operation ........ 43
 6.5.1 Three desludging modes ........ 43
 6.5.2 Two transportation modes ........ 44
 6.5.3 Three scheduling algorithms ........ 44
6.6 Simulate Operation ........ 47

**Step 7**
***Improve institutions*** ........ ***49***

7.1 Assigning Existing Institution ........ 49
7.2 Consider Water Utility ........ 50
7.3 Separate Function to Regulate ........ 52
7.4 The Tasks Define the Structure ........ 53

**Step 8**
***Set regulations*** ........ ***55***

8.1 Must be Mandatory ........ 55
8.2 Many Things to Cover ........ 56

8.3 Regulations are Hierarchical . . . . . 56
8.4 Foster Compliance . . . . . 58
8.5 Promote Regulations . . . . . 58

**Step 9**
***Involve partners*** . . . . . ***61***

9.1 Outsourcing Benefits All . . . . . 61
9.2 Select a Right Partner . . . . . 62
9.3 Agree to Rights and Obligations . . . . . 64

**Step 10**
***Prepare the fleets*** . . . . . ***65***

10.1 No More Handcart . . . . . 65
10.2 Two are Enough . . . . . 67
10.3 Work by Procedures . . . . . 68
10.4 Knowing the Risks . . . . . 70
10.5 Monitor the Operation . . . . . 71

**Step 11**
***Install management system*** . . . . . ***75***

11.1 All is Controlled . . . . . 75
11.2 Rely on Database . . . . . 77
11.3 Connecting the Partners . . . . . 78

**Step 12**
***Plan finance*** . . . . . ***81***

12.1 Each has its Own Tariff . . . . . 81
12.2 Many Payment Options . . . . . 83
12.3 Avoid Financial Loss . . . . . 84

**Step 13**
***Promote service*** . . . . . ***87***

13.1 Prepare Promotion . . . . . 87
13.2 Understand the Target . . . . . 87
13.3 Give a Unique Brand . . . . . 89
13.4 Messages According to Targets . . . . . 90
13.5 Use any Possible Means . . . . . 90
13.6 Involve Other Parties . . . . . 92

**Annexes** . . . . . **93**

# Foreword

Unsafe sanitation is a major problem in our world today, and the cause of many illnesses and deaths every year. This is not a one-country problem, but a global issue. Approximately 4.5 billion people in our world either practice open defecation or use unsafe sanitation facilities and services. There is an urgent need to improve sanitation, and that is why we are excited to make the content in this book more widely available toward that end.

Fecal sludge management is a crucial component of global sanitation services and a strategic solution to addressing unsafe sanitation. It is integral in protecting and improving the lives of the people in communities around the world with non-sewer sanitation systems. Scheduled desludging is a valuable initiative for fecal sludge management systems and provides many benefits for cities and regions that implement it. Scheduled desludging is also an approach that helps ensure inclusive and equitable service delivery so that low-income communities also have access to the services and makes cities accountable for sustainable service delivery. For this reason, the Bill & Melinda Gates Foundation was glad to support the English adaptation of this resource on planning and executing scheduled desludging service which was developed by the great, collaborative, years' of work of USAID Indonesia Urban Water, Sanitation and Hygiene *Penyehatan Lingkungan untuk Semua* (IUWASH Plus) and the Ministry of Public Works in Indonesia.

This book provides helpful and comprehensive guidance to develop scheduled desludging that has been used to establish this service in multiple cities in Indonesia. We look forward to getting this resource into the hands of many people from many countries to learn and benefit from its content. May you, the reader, apply what you find here, and may it serve to support your efforts in improving sanitation services and the many people impacted thereby.

*Roshan Raj Shrestha, PhD*
Deputy Director, Global Growth Opportunity
Water, Sanitation and Hygiene
Bill & Melinda Gates Foundation

# Preface

Over half of Indonesia's 262 million people live in densely populated urban areas that often lack sewerage. In these urban areas, sanitation services are strained; over 80% of this population relies on individual or communal septic tanks that are rarely watertight. Indonesia, however, is highly committed to reaching the UN Sustainable Development Goal 2030 for safely managed sanitation. To improve the health and prosperity of this and future generations of families, Indonesia must increase its self-reliance in providing and managing regular, efficient, and cost-effective collection of sludge from the septic tanks.

From 2011 through 2016, the United States Government, through the United States Agency for International Development (USAID) Indonesia Urban Water, Sanitation and Hygiene (IUWASH) project, assisted the Government of Indonesia to achieve its sanitation development goal through an innovative service for scheduled desludging of fecal waste. This innovative service, called *Layanan Lumpur Tinja Terjadwal* (LLTT) in Indonesian, advances cities' self-reliance in safely handling and disposing of sludge, which is intimately linked with community health, water cleanliness, and fewer instances of waterborne disease such as diarrhea.

Through the LLTT system, households can easily subscribe to a regular septic tank emptying system handled by either public or private sector operators, who transport the sludge to an operating sludge treatment facility. To reduce the burden on families, LLTT facilitates a service payment plan that includes charges in water tariffs or structured monthly payments. This payment plan serves as an incentive for families to subscribe since it makes it easier for them to pay on time and in full.

After three cities launched LLTT, USAID IUWASH developed a book called *Saatnya Sekarang! Jalankan Layanan Lumpur Tinja Terjadwal!* (The Time is Now! Starting Scheduled Desludging!) to promote scheduled desludging and expand LLTT in more municipalities throughout Indonesia. USAID continues to assist 34 municipalities with increasing access to safely managed sanitation services for


doi: 10.2166/9781789062281_xi

families while improving key hygiene behaviors among poor and vulnerable urban populations. In early 2020, IUWASH PLUS updated and refined the technical, financial, and institutional aspects of the LLTT approach, and to date, 13 of the 34 assisted cities have launched LLTT services. With the support of USAID and the Ministry of Public Works, Ministry of National Planning, and other donor agencies, scheduled desludging of fecal waste is now happening in many other cities throughout the country, meaning that more families can enjoy better health and hygiene.

USAID thanks the Bill and Melinda Gates Foundation for helping to make the English-language version of this book a reality.

*Matthew Burton*
Environment Office Director
USAID Indonesia

# Introduction

Almost all countries have committed to achieve sustainable development goals (SDG) by 2030, one of the goals is to achieve the provision of safety managed sanitation for all citizens. This means that each resident must have access to a toilet that is connected to a sewer system or to a septic tank. Many cities do not have sufficient funds to build and operate sewerage systems, consequently they encourage every household to use proper septic tank.

Septic tanks accumulate solids, either carried by the incoming wastewater or resulting from the anaerobic decomposition of feces and other organic materials. The settled solids or sludge must be pumped out regularly, otherwise the septic tank efficiency will be disrupted and septage might pollute the ground water. The accumulated solids can also reduce or even stop the flow of wastewater and disrupt the function of the toilet. Considering the potential negative impacts, periodic desludging of septic tanks has been made mandatory in cities in several countries, including in Indonesian cities.

A city or municipality must provide a service scheme that can provide periodic desludging to households in accordance with a specified schedule. Several programs need to be in place so the scheduled desludging scheme can run effectively, including regulations, institutions, infrastructure and customer management. In addition to service providers, municipal agencies need to be involved to ensure households use the right septic tank and perform regular desludging.

This book presents a methodology for preparing scheduled desludging schemes in cities. Opening with a description of its definition and principles, this book outlines step by step the preparation of a scheduled desludging scheme. Each city has its own characteristics; therefore, it is possible that a city may have a different sequence of preparation steps than other cities.

This book is an international adaptation from the book of *Saatnya Sekarang! Jalankan Layanan Lumpur Tinja Terjadwal (It's Time Now! Starting Scheduled Desludging Service)* which was created and published in 2015 by the Indonesian Urban Water, Sanitation and Hygiene (IUWASH) program funded by the United States Agency for International Development (USAID). This book targets those who will be involved by the municipalities to prepare scheduled desludging schemes in their respective regions. Some of Indonesia's specific narratives and terms used in the original book have been adapted for the benefit of readers from other countries (see the following box).

Happy reading.

## The original book

*The book of Saatnya Sekarang! Jalankan Layanan Lumpur Tinja Terjadwal (It's Time Now! Run the Scheduled Desludging Service) was prepared in 2015 as part of IUWASH (Indonesian Urban Water, Sanitation and Hygiene) program activities to promote the scheduled desludging scheme in Indonesia. This book outlines the steps that need to be taken in preparing scheduled desludging in a city.*

*Much of the materials in this book are taken from IUWASH program's direct experience in the introduction and preparation of scheduled desludging in several Indonesian cities, for example, Makassar, Surakarta (Solo), as well as DKI Jakarta. Some materials are taken from the Guidelines of Septage Management published by the Directorate General of Human Settlements (the Ministry of Public Works and Public Housing), as well as from the reports published by the Water and Sanitation Program (World Bank).*

*This book is complemented by 4 supplementary books on K3, finance, marketing and operations. The authors are IUWASH's internal expert staff in these fields and several external consultants. Periodic consultation is carried out with representatives of the Indonesian government, especially those from BAPPENAS (National Development Planning Agency) and the Ministry of Public Works and Housing. Soft copies of these books can be downloaded from the official website of the IUWASH PLUS program and other sites. The USAID-funded IUWASH PLUS program will continue to work in Indonesia until end-2021.*

*Adaptation for this book was prepared by Rudy Yuwono, Stantec Consulting Services, Inc., and Asian Institute of Technology (AIT).*

# Understand scheduled desludging

*We must understand the definitions, roles, benefits, and aspects of scheduled desludging before we start preparing them in our city. It must be clearly understood that scheduled desludging is a measure given in response to the obligation for septic tanks to undergo periodic desludging. Scheduled desludging is a component of the septage management of the city; its effectiveness, performance and sustainability are greatly influenced by the other components. This section concludes with a discussion of the steps for preparing a scheduled desludging scheme in the city.*

## 0.1 A MANDATORY DESLUDGING

Septic tanks have the ability to digest organic material and separate floating material and solids from domestic wastewater. Settled solids are digested anaerobically while the liquid portion overflows into the soak pits or infiltration gallery. Settled solids need to be removed from the septic tank on a regular basis to prevent it from accumulating and reducing the capacity of the tank. Regular desludging will keep the septic tank functioning properly. Once every 2–3 years is recommended as long as the tank is used by the same number of people under the design assumption.

The scheduled desludging is the periodic desludging of the septic tank as required by regulation (see Figure 0.1 and **Step 8: Set regulations**). The municipality, through their designated service provider, determines the desludging time of each septic tank in the area. Thus, septic tanks under a scheduled desludging scheme are not emptied due to requests from households. The desludging is carried out according to the specified schedule. The scheduled desludging is usually conducted every 2–5 years (see Figure 0.2), although the exact frequency will be determined after the municipality comprehends the general characteristics of the septic tanks in their area (see **Step 6: Design operations**).

The scheduled desludging is managed by an organization appointed by the municipality (see **Step 7: Improve institutions**). If necessary, the designated service provider can engage a private desludging company to assist with the task (see **Step 9: Involve partners**).


doi: 10.2166/9781789062281_0001

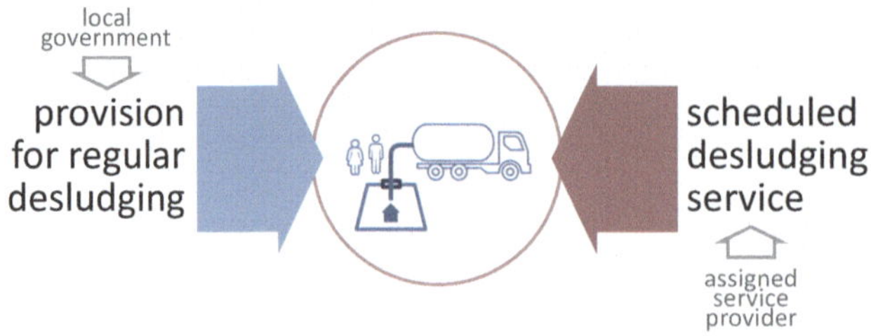


**Figure 0.1** Scheduled desludging service is provided in response to regulatory provision requiring septic tanks to be regularly desludged. All septic tank users by law must accept the service.

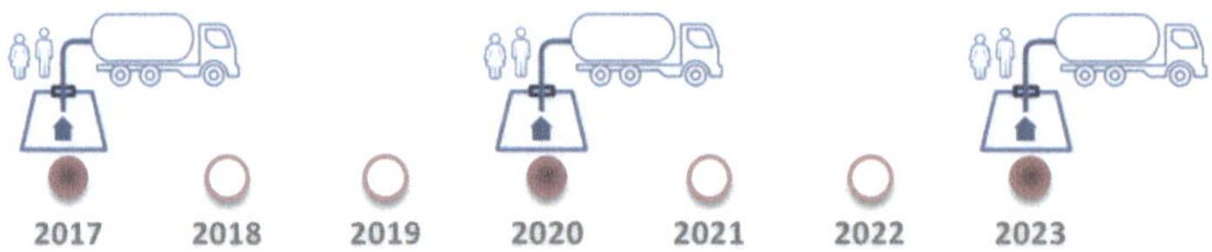


**Figure 0.2** The desludging period is the time span between consecutive desludgings. A 3-year desludging period means a septic tank will be emptied in 2017, 2020, 2023, and beyond.

**Figure 0.3** The City of Surakarta (Central Java) has implemented the scheduled desludging scheme as required by the local regulations. Each septic tank must be desludged every 3 years. The local water supply firm is appointed to manage the scheduled desludging in 2015.

Scheduled desludging fits well with citywide inclusive sanitation (CWIS). This approach requires guarantees that the services provided by the municipalities will be utilized so that all residents benefit from the existing sanitation system. The existence of continuous and periodic services from scheduled desludging provides onsite solution effectiveness which is almost as good as offsite solution, particularly in handling fecal waste.

## 0.2 DESLUDGE AND TRANSPORT

A septage (or, fecal sludge) management system consists of four components: (1) septic tank control or supervision, (2) septic tank desludging, (3) septage transportation and (4) septage treatment (see Figure 0.4). Separation of fecal management components can be carried out by cities that want to assign different service providers or establish separate budgets for each of the above components. The scheduled desludging scheme covers only two components, namely septic tank desludging and septage transportation. The septage treatment in several cities is typically managed by a different organization than the septic tank desludging. They also receive and treat septage from on-call or on-demand desludging. Unlike scheduled desludging, on-demand desludging is only done if there is a request from the household (see the following table for the differences between the two types of desludging).

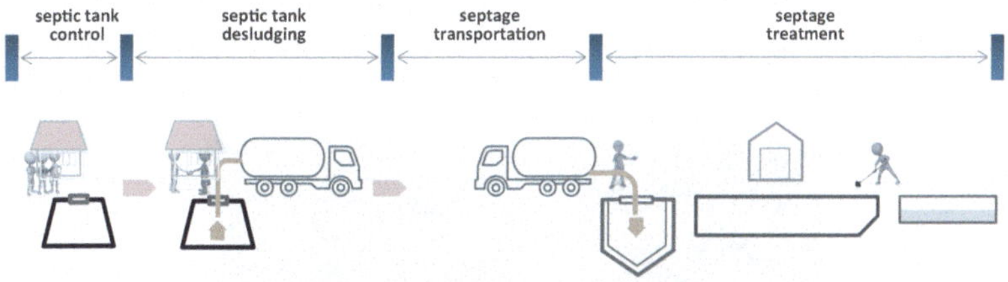


**Figure 0.4** In the septage management chain, septic tank desludging occurs between septic tank control and septage transportation to treatment. The performance of each component is very important for the overall continuity and performance of the septage management system.

Scheduled desludging does not replace on-demand desludging. Both types must be available side-by-side in the city because there are always households that need more frequent or emergency septic tank emptying. The scheduled desludging must be regulated institutionally while on-demand desludging does not always require any arrangement from the municipality.

Scheduled desludging vs on-demand desludging.

| | Scheduled Desludging | On-demand Desludging |
|---|---|---|
| Nature | Mandatory | Voluntary |
| Implementation | Scheduled by the municipality | Requested by the households |
| Customers | Registered | — |
| Institutional | Requires a citywide managing service provider | — |

## 0.3 THE MANY BENEFITS

Scheduled desludging has direct and indirect benefits. The direct benefits of scheduled desludging are:

- improve performance of septic tanks in the city,
- maintain continuous supply of septage to treatment facility,
- reduce environmental pollution and improve public health,
- provide higher revenues.

The indirect benefits of scheduled desludging are:

- raise public awareness about their responsibilities in managing wastewater, including its financial implication,
- create business opportunities for private sectors,
- enhance image of a city.

It should be noted that the above benefits will only occur if all components of septage management function properly. In a city with many sub-standard septic tanks, scheduled desludging will not optimally prevent pollution. However, the scheduled desludging operation with its regular house visits can assist the municipality to identify and record sub-standard septic tanks (see Figure 0.5). Scheduled desludging will not bring optimal benefits if a city does not have a well-performed septage treatment plant.

It should also be noted that scheduled desludging is not an easy scheme to implement. Some cities fail to establish it, others fail in maintaining its sustainability. The introduction of scheduled desludging gave new

**Figure 0.5** Scheduled desludging ensures all septic tanks in the city will be periodically inspected. The municipality will have data on the existence and condition of septic tanks in the area and the government will have a basis to enforce their septic tank regulations to the households.

obligations to households so it was often met with rejection. In implementing cities, households are not always willing to accept the desludging services that have arrived in the location. This also happens even though the households have actually paid the monthly service fee. Public awareness raising is needed before scheduled desludging is developed in a city. Due to the various conditions of the septic tanks, scheduled desludging operators often find it difficult to carry out their duties, especially in the first cycle of scheduled desludging.

## 0.4 SEVEN ASPECTS TO CONSIDER

Scheduled desludging has seven aspects that need to be considered in its preparation and implementation, these include the operations, customers, infrastructure, regulations, institutions, procedures, and finances (see Figure 0.6). These seven aspects form an integrated joint support system to ensure scheduled desludging operations are carried out systematically, properly and sustainably. If one aspect is ignored, it is likely the scheduled desludging will not function as expected.

The following is a description of each aspect of scheduled desludging preparation and implementation.

(1) **Operation scheme:** Scheduled desludging must have an appropriate operating scheme in accordance with regional conditions and service targets, specifically regarding (a) the desludging period, (b) service zonings, (c) desludging volume, (d) scheduling algorithm and (e) transportation routes.

(2) **Customer:** Scheduled desludging must have enough customers to optimize their service operations and bring in large financial revenues. Scheduled desludging customers must meet the following criteria: (a) septic tank user, (b) location accessible by desludging trucks, and (c) willingness to pay for services.

(3) **Infrastructure:** Scheduled desludging needs to be supported by desludging and transportation vehicles, treatment facility, and office and management information systems. All infrastructure must be selected and provided in accordance with the operating scheme and financial condition of the city or service provider.

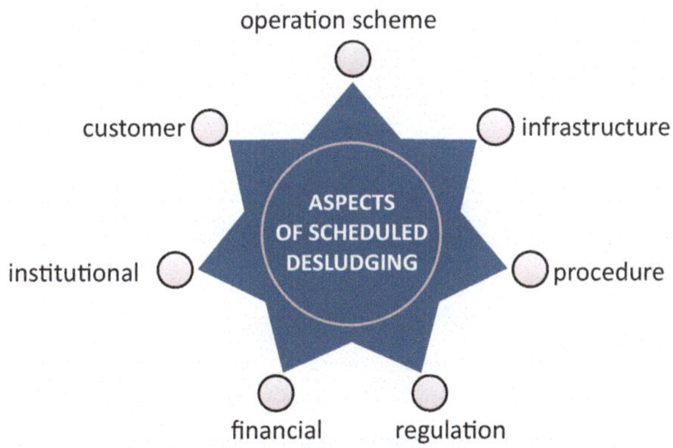


**Figure 0.6** Scheduled desludging has seven management aspects to allow desludging to be carried out properly, systematically and sustainably. The seven aspects must be developed in accordance with the characteristics and capabilities of the city, as well as with the service objectives and targets to be achieved.

(4) **Institutional:** The performance and sustainability of scheduled desludging needs to be supported by institutions that have specific functions, namely planning, compliance with regulations, operations management, and service delivery. One organization may be able to carry out several of these functions as long as it does not create a conflict of interest. Scheduled desludging can involve private companies to carry out the service delivery function.

(5) **Procedure:** Scheduled desludging must be supported by (a) a customer management procedure, (b) a septic tank desludging procedure, (c) a septage transportation procedure, (d) a customer billing procedure, (e) a performance evaluation procedure. The consistent implementation of operating procedures will make scheduled desludging run regularly and systematically in accordance with the agreed operating scheme and objectives.

(6) **Financial:** Scheduled desludging must earn enough revenue from the households to cover all operating and management costs. To the extent possible, revenues from the scheduled desludging service can be used to help finance the infrastructure investment and provide a reasonable profit to the municipality or the service provider.

(7) **Regulation:** Scheduled desludging needs to be supported by a set of regulations requiring: (a) proper specifications and use of septic tanks, (b) periodic desludging, (c) monitored septage transportation, (d) disposal of septage in treatment facilities or other designated points, (e) septage treatment to meet effluent and other environmental standards, and (f) payment of service rates. In addition, regulations must also provide direction on (a) institutional framework, (b) private involvement, (c) payment mechanisms, and (d) tariff of the service.

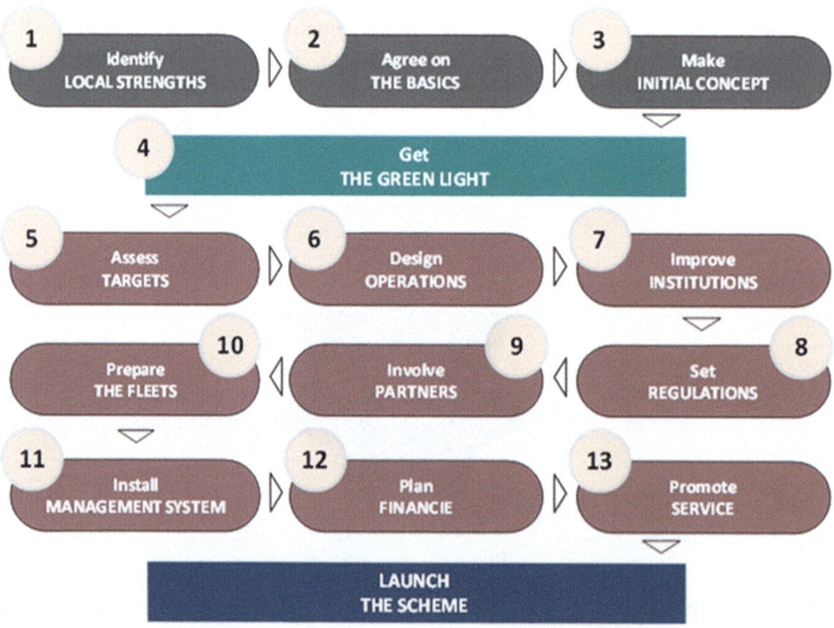


**Figure 0.7** The 13 generic steps to set up the scheduled desludging scheme in a city.

## 0.5 THIRTEEN DEVELOPMENT STEPS

There are 13 general steps for preparing or setting up a scheduled desludging scheme in the city (see Figure 0.7). Some steps are sequential, others can be done simultaneously. Each city has its own characteristics and abilities so the steps in one city may be different from those in another city. Before the first step, the city should form a team that will facilitate the development of a scheduled desludging scheme. The following are the 13 steps to set up a scheme (see diagram).

(1) **Identify local strengths:** The team should obtain information on the population of the community and building use, the level of use of septic tanks, the availability of desludging trucks, the capacity of septage treatment facilities, and the existence of regulations and institutions related to sanitation issues. The team should investigate which of these items allow setting up of the scheduled desludging scheme to be feasible and less complicated and which of these will be the bottleneck.

(2) **Agree on the basics:** There are a number of basic and fundamental issues that need to be discussed and agreed on between team members and relevant parties that will provide general direction to the preparation of the scheduled desludging scheme. This might include the targets (area, types of buildings, and level of coverage), timelines, stages of development, and financial principles.

(3) **Make initial concept:** The initial concept should at least describe the estimated number of households to be served, the scale of operations, required infrastructure and financial estimates. The initial concept is made using existing secondary data and several logical assumptions.

(4) **Get the green light:** The green light from city leaders, particularly the mayor, must be obtained before implementing a scheduled desludging scheme. The green light of the leadership will enable the team to get support from other agencies. The initial concept of the scheduled desludging scheme must be presented to the leaders

(5) **Assess targets:** The scheduled desludging scheme is intended for all households and other buildings that use accessible septic tanks. The team must collect information from all households and buildings before a building can be classified as a viable scheduled desludging target or customer. At the end of this step, the team should have information on the amount, spatial distribution and type of buildings as well as the specifications and conditions of a proper septic tank. The best way is to survey or to perform a census on all households and buildings in the city.

(6) **Design operations:** Information collected from surveys or a census is used as a basis for determining the service zones, amount and classification of targeted buildings, desludging periods, scheduling period, and desludging modes. The operations design also includes the daily volume of septage, desludging frequency, number of desludging trucks or other vehicles and required treatment capacity for septage originating from scheduled desludging operation.

(7) **Improve institutions:** After the main service provider of scheduled desludging is appointed, the team should plan the capacity building activities for the organization. Starting from setting the organizational structure, improving functions and increasing the number and qualifications of personnel. In addition to service providers, the city government also needs to determine the role of planners, regulators and supervisors in other institutions. All must have the capacity needed to support scheduled sludge removal schemes. In addition to the service provider, the municipality needs to determine and establish the role of a planner, regulator, and supervisor on the other institutions. These roles must have the capacity required to support a sustainable and well-regulated scheduled desludging scheme.

(8) **Set regulations:** There must be regulations to require the use of proper septic tanks which must be desludged regularly and periodically as well as requirements for transporting septage safely to the

treatment facility where the septage will be treated to meet the effluent and other environmental standards. Regulations regarding service tariffs in several cities require approval from the legislature. All regulations must be developed together with their community promotion and compliance strategies.

(9) **Involve partners:** Third parties, both private entities and community groups, may need to be involved in providing septic tank desludging services. After choosing the right partner, we need to fix the scope and form of cooperation between the main service providers and the partners. Agreement must be made in writing and officially signed.

(10) **Prepare the fleets:** The septage desludging fleet needs to be prepared according to the operating scheme, modes of desludging and transportation and general conditions of traffic. This includes desludging vehicles, crew members and operating procedures. A licensing scheme for desludging vehicles needs to be established, as well as a fleet monitoring system. All crew members must undergo trainings before they can carry out their duties.

(11) **Install management system:** The quality of the business process of the scheduled desludging service provider will be determined by the quality of the management system. By utilizing digital information technology, we can manage information consistently, accurately, and efficiently. The management system will also be connected to a fleet monitoring device so that all operations can be monitored and recorded in real time.

(12) **Plan finance:** We need to estimate capital and operational expenditure based on the agreed operating design. Tariffs can be calculated afterwards according to the cost recovery policy and customer groupings. Projections on profit and loss, as well as balance sheet should be prepared to better ensure the financial health of scheduled desludging operation. The team can adjust the operational plan until the financial plan meets the expectations of all parties.

(13) **Promote the service:** Scheduled desludging is a mandatory measure. All households must understand the reasons and benefits of scheduled desludging as well as their rights and obligations as customers. A promotion strategy must be developed, before marketing tools and promotional officers are prepared.

The steps above are open for modification according to the characteristics and capabilities of each city. The speed at which scheduled desludging is prepared will be affected by the intensity and effectiveness of the team and all parties involved. It is also possible to adopt a more conservative approach that starts with raising awareness of on-demand emptying, applying incentives to those who request regular service before going for full-fledged scheduled desludging

# Step 1

# Identify local strengths

*Begin the process of preparing a scheduled desludging scheme by identifying things that can make the process simpler and faster. In the process, obstacles may be encountered that can slow down or even halt the process. Assess the current conditions of the seven scheduled desludging aspects against the conditions required. At the end of this step, we must decide whether preparation of scheduled desludging is feasible to continue and determine what measures are needed to close the gap.*

## 1.1 ANALYZE THE GAP

Initiate the development of a scheduled desludging scheme by finding answers to the following questions: what is currently owned by a city that can accelerate the preparation process of the scheme? What will halt or slow down the preparation of the scheme? To answer these questions, the team of course needs to understand what aspects of the scheduled desludging scheme are needed so it can be implemented properly and sustainably (see Figure 1.1). Not only the technical aspects, but also the institutional, regulatory and financial aspects. The team will later compare what the city already has with what is needed for a scheduled desludging scheme, also called a gap analysis. Local strengths should be identified as well as the bottlenecks.

## 1.2 RELY ON AVAILABLE DATA

In this step, the team can utilize existing secondary information or data available at municipal agencies related to urban planning, infrastructure provision, building control, housing and settlements, public health and sanitation services. This step does not include going to the field to get primary data and information. Information to be collected includes:

- the total population and buildings,
- the use of septic tanks and their general conditions,


doi: 10.2166/9781789062281_0009

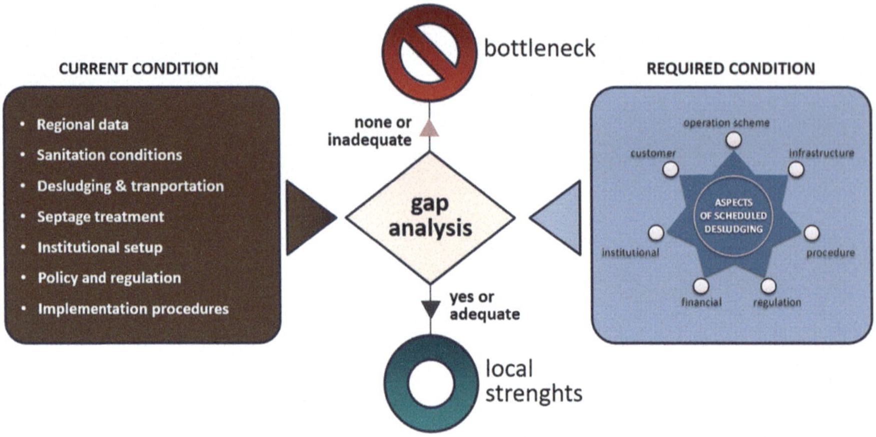


**Figure 1.1** The local strengths or bottlenecks for preparing a scheduled desludging scheme are identified after we compare the current conditions with the required conditions of a proper and sustainable scheduled desludging scheme. At the beginning of the preparation process for scheduled desludging, gap analysis can be done using only secondary information.

- the availability and coverage of wastewater sewerage system,
- the availability of desludging services,
- the number of trucks or other desludging units,
- the availability, capacity, and condition of septage treatment and other wastewater treatment facilities,
- the institutions related to wastewater and septage managements, as well as to water supply, cleansing, and drainage,
- the regulations related to wastewater, septage, buildings, and environmental managements

As much as possible, any information obtained should include its source along with year and location. It is helpful to use a questionnaire or checklist to guide us in gathering information (see Figure 1.2 and **Annex A** for the questionnaire).

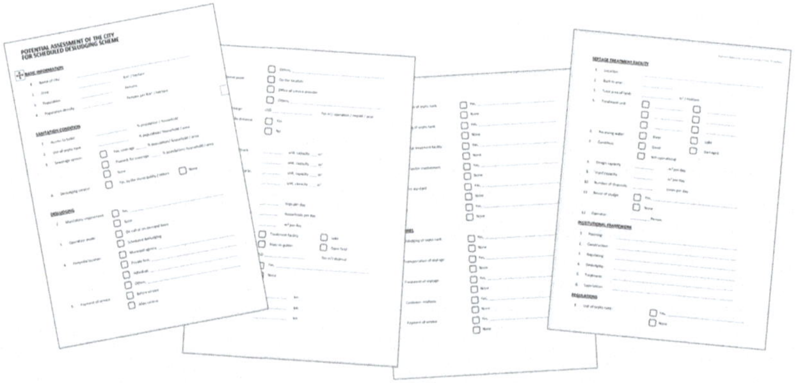

**Figure 1.2** The Assessment Form of City Strengths for Scheduled Desludging used by IUWASH PLUS program officers and their counterparts when assessing the conditions of the city where they work to develop scheduled sludge removal services.

## 1.3 'GO' OR 'NO GO'

The team should evaluate whether a city is ready to develop scheduled desludging scheme based on information collected. There is a possibility that at the end of this step the team concludes the city is not ready for a scheduled desludging scheme. Reasons may include the small number of households that use septic tanks or due to the absence of septage treatment facility in the city (see Figure 1.3). If these two problems are not found, the city can be considered feasible to immediately start the development of a scheduled desludging scheme.

A key issue that prevents a city from having a scheduled desludging scheme is a small number of septic tanks within its area due to the restricted capability of the scheme becoming financially independent. The lower the number of septic tanks, the lower the economic scale of the scheduled desludging scheme. It is a fact that many cities still have low levels of access to improved sanitation, which means the toilets do not connect to a septic tank or to a sewerage system network. Some cities even still have high open defecation rates. In such cities, it is far better for the municipalities to optimize on-demand desludging services while also promoting the proper use of septic tanks or sewer connection.

Another main issue is the existence and capacity of a septage treatment facility (see Figure 1.4). Many cities do not have such treatment facilities, or even access to a treatment facility in the nearest city. Meanwhile some other cities may have septage treatment facilities, but these facilities may not be in good condition or even really damaged. A city must at least be able to show plans to develop their septage treatment facility in order to be deemed feasible to start preparing the scheduled desludging scheme.

Another issue is the domestic wastewater management regulation which many cities do not have. Cities may have regulations regarding buildings, environment management or public health which contain a number of provisions regarding wastewater management. However, these regulations rarely contain detailed provisions to support an effective scheduled desludging scheme; for example, the requirements

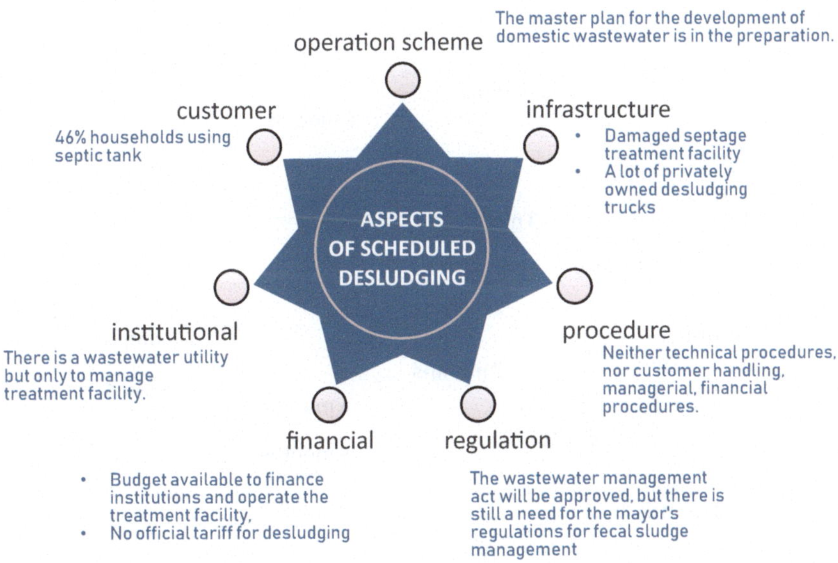


**Figure 1.3** A brief note of the result of the gap analysis of city potential carried out by IUWASH team in an Indonesia city. At the end of the analysis, the city was deemed fit to start preparing a scheduled desludging scheme provided they had a plan to rehabilitate their treatment facility and improve the regulations.

**Figure 1.4** Septage treatment facility is an important component of scheduled desludging schemes. Their capacity often becomes the limiting factor that determines the coverage of septage desludging scheme in the early years.

to have database on septic tanks, 3-year desludging time, monitored desludging and transportation, safety gears, monthly tariff payment and others may not be provided in detail. A specific septage management regulation is a must for a city to implement a scheduled desludging scheme.

The team can make a list of all the follow-up measures necessary to develop a scheduled desludging scheme in the city. Arrange the steps in sequence in a diagram to make it easier for others to understand (see Figure 1.5). Enter the estimated completion time for each measure so we can calculate the total time needed to prepare a complete scheduled desludging scheme in the city.

The preparation of a scheduled desludging scheme for some cities could be their first attempt in preparing an institutionalized citywide sanitation service. A lot of resources, energy and time will be spent by the municipalities to set up the supporting components of the scheduled desludging scheme. These components will also be needed if the city wants to develop a sewerage system service in the future.

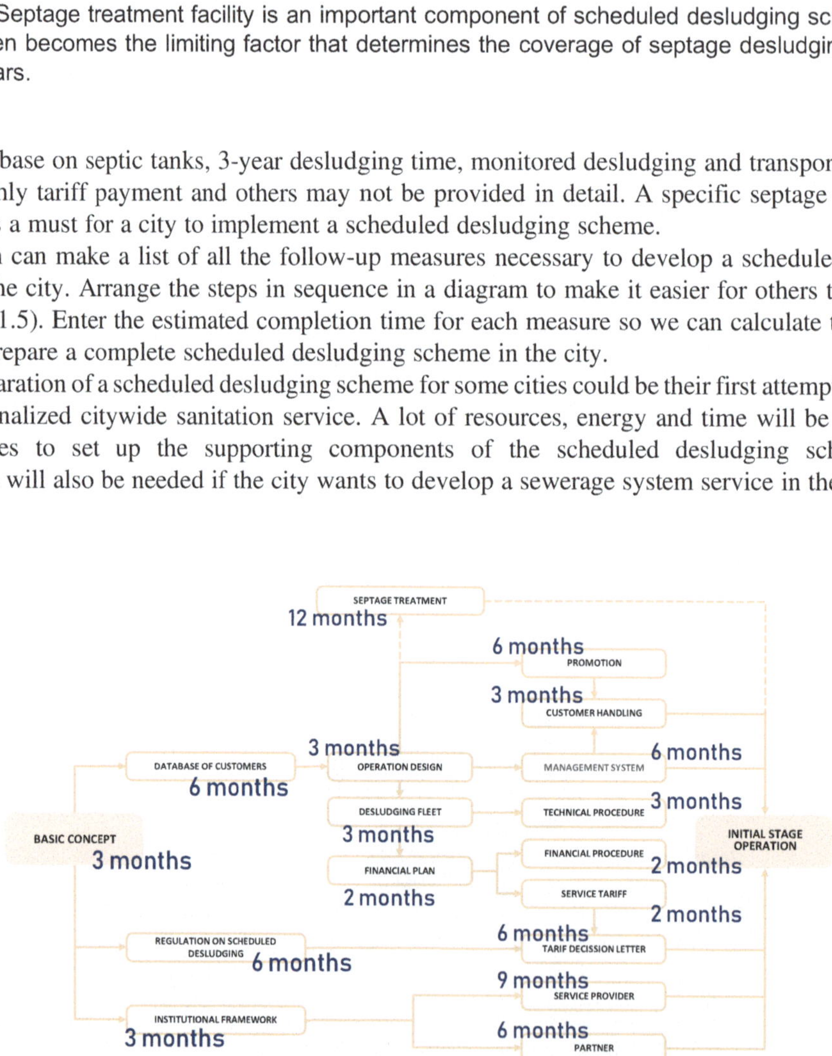


**Figure 1.5** Flow chart of measures for developing a scheduled sludge desludging scheme in the city as a result of an assessment of the city's potential. The duration for carrying out each measure needs to be specified so the team can estimate the total time needed to prepare a scheduled sludge removal scheme in the city.

# Step 2

# Agree on the basics

*There are some basic decisions that need to be agreed upon by related parties which will drive the preparation process. These basics may include the targeted areas and level of coverage, development stages, timeline and financial costing principles. It is important to involve representatives from other agencies who later will be involved in the preparation or the implementation of the scheduled desludging. This step will only be effective if all parties have the same perception about the scheduled desludging scheme.*

## 2.1 ESTABLISH COMMON PERCEPTION

Preparation of a scheduled desludging scheme in the city must involve representatives from various agencies who inevitably, directly or indirectly, will be involved in the operations and supervision of the scheme. This includes agencies of urban planning, infrastructure provision, building control, housing and settlements, sanitation, public health, and environmental management.

A workshop must be held to ensure all parties have the same understanding and perception of a scheduled desludging scheme. A few examples of topics to discuss in the workshop include:

- Scheduled desludging includes a mandatory septic tank emptying to be conducted periodically as required by law to all septic tanks.
- Scheduled desludging is conducted in a period and time set by the municipalities (with input from service provider).
- Scheduled desludging only handles septage which is defined as all mixtures of solids and liquids including human waste that accumulates in the septic tank.
- Scheduled desludging service consists of septage desludging and transportation activities while septage treatment is a different service which may be conducted by another service provider.
- Scheduled desludging does not eliminate the existing on-demand desludging service (see Figure 2.1).
- The service provider assigned to manage scheduled desludging operations is not automatically assigned to regulate on-demand desludging service operations unless the regulation orders it.


doi: 10.2166/9781789062281_0013

**Figure 2.1** On-call or on-demand services will not be eliminated due to the establishment of a scheduled desludging scheme in the city. Both will be complementary to each other because many households may require more services than the time period set for scheduled desludging. https://sedot-wc-bandung-juara.business.site/.

## 2.2 DEFINE OUTCOMES

We can start this step by defining the outcomes of the scheduled desludging scheme in the city. An outcome is a long-term result due to the existence of the scheduled desludging scheme. For each outcome, it is important to understand the implications and to set corresponding indicators (see the following table).

Examples of outcomes.

| Outcomes | Indicators |
|---|---|
| Improved overall sanitation behavior and services | • The level of use of septic tanks<br>• Amount of fecal sludge received by septage treatment plant |
| Improved quality of the environment | Percentage of groundwater samples contaminated with *Escherichia coli* |
| Improved level of public health | Number of cases of illness due to poor sanitation conditions |
| Improved image of the city | Number of acknowledgments or awards from other parties |

## 2.3 SET AREA AND TIME

Two basic things that need to be agreed on from the beginning of the initial step of scheduled desludging preparation are (a) the area where scheduled desludging will be applied and (b) the time when the scheduled desludging will start. A master plan for the development of domestic wastewater system or a city sanitation strategy may contain information relating to both (see Figure 2.2). In cases where a city does not have such documents, the team can seek input and recommendations from urban planning and environmental management planning agencies.

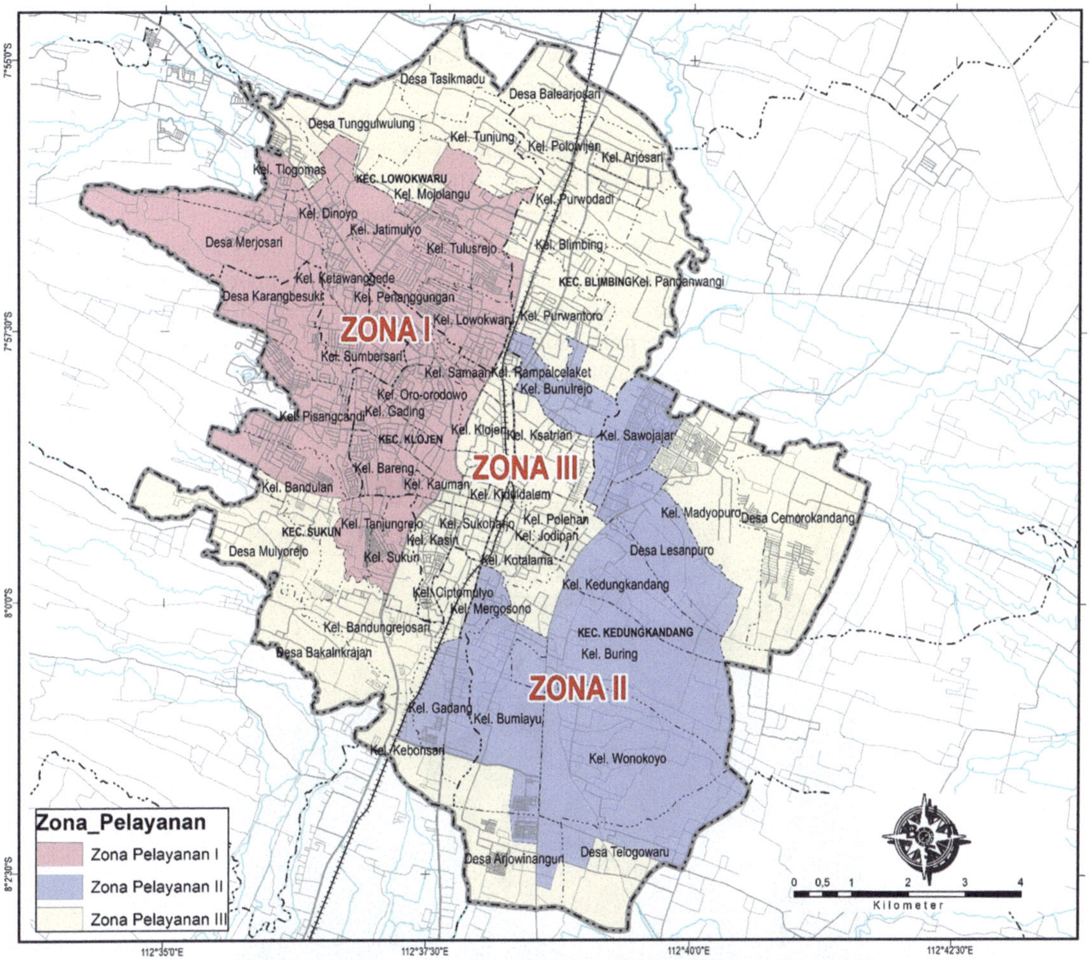


**Figure 2.2** The master plan for domestic wastewater management of a city needs to be considered in determining scheduled desludging target areas. The map above is taken from the Malang city wastewater master plan (Indonesia) which shows three wastewater service zones to be developed in the city. Zones I and II are planned to have a wastewater sewerage system while zone III will rely on an on-site system. The scheduled desludging will be the main service in zone III.

Scheduled desludging will be required for all septic tanks in the city. However, there are certain conditions where a scheduled desludging scheme is not suitable for some parts of the city. These conditions may include (a) areas with high open defecation rates, (b) areas with unclear soil status, or (c) areas with existing and future sewerage systems. A scheduled desludging scheme can start at any time according to decision of the city government. However, it is important to ensure the scheduled desludging scheme is started when the infrastructure is ready, and the gap has been resolved (see the section in **Step 1: Identify local strengths**). Depending on the level of preparedness of the scheduled desludging aspects, it may be an advantage to introduce a pilot trial phase in the first year of introduction.

## 2.4 PROPOSE INSTITUTIONS

Scheduled desludging schemes require the involvement of at least two types of institutions, that is (a) the regulator and (b) the service provider(s). Some cities may already have institutions able to carry out these roles. However, the scheduled desludging scheme may be the first city-wide service that the city will have. Existing institutions may not have sufficient authority or capacity to carry out such large-scale services. The gap analysis conducted in the previous step will provide us with information regarding the capacity of the existing institutions in the city (see **Step 1: Identify local strengths**).

The team can identify and propose an institutional setup for the scheduled desludging scheme, but it will be the city government who will make the final decision. However, it is very important to indicate which institutions or organizations will play the roles of regulator and service provider from the start (see Figure 2.3). They must be involved in the preparation of the scheduled desludging scheme. They can provide input and suggestions about their future roles, as well as the required capacity building activities.

**Figure 2.3** The role of regulators and service providers cannot be held by one institution. There are regulatory and control functions that need to be carried out by the regulatory body, while service providers do require supervision so that operations can proceed according to standard procedures.

## 2.5 RECOMMEND FINANCIAL PRINCIPLE

The financial principles of STP operations should be agreed in this step. This is needed to calculate the amount of costs that will be borne by households (conducted in the next step). One choice is to burden households with sufficient costs to recover all operational and maintenance costs (or, an operational cost recovery structure). Another option, burden households with all costs and investment fund of future infrastructure (or, a full cost recovery structure). If scheduled desludging is expected to be profitable, of course the burden of costs to households will be even higher. However, there is a possibility that the local government will need to provide subsidies to ease the burden on the households in the service area.

# *Step 3*

# Make initial concept

*The team can make the initial concept of the scheduled desludging scheme to be presented to the city government and other key stakeholders in the city. It is important they have a good understanding of the service scheme that will be proposed for their city. The initial concept should be developed in accordance with the basics previously agreed upon. In addition to the number of households to be served and the required infrastructure, the initial concept needs also to show financial estimates. The team can use secondary data and common sense to formulate the initial concept.*

## 3.1 JUST AN ILLUSTRATION

The initial concept of a scheduled desludging scheme must show, at a minimum:

- the number of households to be served,
- the daily desludging frequency and septage volume,
- the required number of required desludging units and the capacity of the septage treatment plant,
- the desludging tariff and annual operating costs and revenues.

The initial concept should cover the portion of the city selected to be served in the first year of the scheduled desludging scheme (see **Step 2: Agree on the basics**).

The team may not have actual information on the number, size, condition, or accessibility of the septic tanks in the selected area, which would be needed for a detailed design of a scheduled desludging scheme of a city, but these are not necessary to put together an initial concept of the scheduled desludging scheme. The team can use the secondary information partially obtained in the previous step (see **Step 1: Identify local strengths**). This is still acceptable because now we only want to make an illustration of the future scheduled desludging scheme. Such illustration will be a helpful tool to introduce the scheme to the city government, other city leaders, and stakeholders.


doi: 10.2166/9781789062281_0017

The initial concept is based only on the number of households that use septic tanks. Other types of buildings are not included in this step, that is, public offices, commercial buildings, and social buildings. This is done solely for simplification, especially in the early stages of the preparation. Later, the team will elaborate the scheduled desludging concept by including all types of buildings in the city.

## 3.2 ESTIMATE OPERATIONS

The initial concept of a scheduled desludging scheme, as mentioned earlier, presents the estimated number of households to be served, frequency of desludging, as well as the number of desludging trucks and capacity of septage treatment. These can be estimated from simple calculations using secondary information and common assumptions. The following are examples of calculations to estimate the operations:

(1) Obtain information on the population and households:

| ***Parameter*** | ***Example*** |
|---|---|
| • *Population (people)* | *400,000 people* |
| • *Average number of people per household (people/household)* | *5 people/household* |
| • *Proportion of septic tank users (% number of households)* | *80% of households* |

(2) Assume and agree on the values of the following operational parameter.

| ***Parameter*** | ***Example*** |
|---|---|
| • *Tank volume of the desludging truck* ($m^3$) | *3.0* $m^3$ |
| • *Number of operation days (days/year)* | *250 days/year* |
| • *Working hours (hours/day)* | *8 hours/day* |
| • *Average desludging operation time (hour/household)* | *0.5 hour/household* |
| • *Average trip time to the septage treatment plant (hour/trip)* | *0.5 hour/trip* |
| • *Desludging period (year)* | *3 year* |
| • *Service coverage (% households using septic tanks)* | *75% of households using septic tanks will be served* |
| • *Average desludging volume* ($m^3$*/household)* | *1.5* $m^3$*/household* |

(3) Calculate the number of served households:

| ***Parameter*** | ***Example of Calculation*** |
|---|---|
| • *Number of served households (households)* | *(400.000 people)/(5 people/household) × (80%) × (75%) =* *48.000 households* |
| • *Number of served households per day (household/day)* | *(48.000 household)/(3 years)/(250 days/year) =* *64 household/day* |

(4) Calculate the number of operation cycle:

| Parameter | Example of Calculation |
|---|---|
| • *Capacity of septage in 1 operational cycle ($m^3$/cycle)* | *3 $m^3$/cycle* |
| • *Number of served households in 1 operational cycle (household/cycle)* | *(3 $m^3$/cycle)/(1.5 $m^3$/household) = 2 households/cycle* |
| • *Number of operational cycles per day* | *(64 households/day)/(2 households/cycle) = 32 cycles/day* |

(5) Calculate the number of desludging trucks required:

| Parameter | Example of Calculation |
|---|---|
| • *Time required for one operational cycle (hour/cycle/truck)* | *[(2 households/cycle) × (0.5 hours/household)] + [(2 trips/cycle) × (0.5 hours/trip)] = 2 hours/cycle* |
| • *Number of operational cycles per desludging truck (cycle/day/truck)* | *(8 hours/day)/(2 hours/cycle/truck) = 4 cycles/truck/day* |
| • *Number of desludging truck required (truck)* | *(32 cycles/day)/(4 cycles/truck/day) = 8 trucks* |

(6) Calculate the required capacity of septage treatment:

| Parameter | Example of Calculation |
|---|---|
| • *Volume of septage that needs to be treated ($m^3$/day)* | *(64 households/day) × (1.5 $m^3$/household) = 96 $m^3$/day* |

We can use spreadsheet software to speed up the calculation process above, such as Microsoft Excel.

## 3.3 CALCULATE BASIC COST

In the context of this book, basic cost means the cost to deliver a sustainable scheduled desludging service in order to set household tariffs while considering profit and loss. The basic cost is the same as the average tariff to be imposed to households provided the cost recovery principle is applied. Collectively, the total revenue from the households will fully finance scheduled desludging operations, including government subsidies.

The calculation of the basic cost of scheduled desludging operation should include all operating costs which are grouped as follows:

- **Collection cost:** The cost required to desludge septic tanks and transport the septage to the treatment plant, including fuel costs, maintenance costs, vehicle tax, honorarium, and communication costs in the calculation. The existing service provider in the area may have insights to incorporate into the collection cost.

- **Management cost:** The cost required for wages or salaries, office overhead, promotions, as well as for the depreciation of assets. The amount of the management cost is also influenced by the scale of scheduled desludging operation.
- **Treatment cost:** The fee required to pay the treatment service for each septage disposal at the treatment plant. The treatment fee is usually already set by the septage treatment service provider or by the municipality. The items in the treatment cost typically consist of energy (electricity and fuel) cost, material cost, maintenance cost and operator cost. The amount of the treatment cost is strongly influenced by the type of technology and by the treatment capacity.

The basic cost can be calculated and presented as a monthly cost using the following formula:

$$\text{Basic cost} = \frac{(\text{collection cost}) + (\text{management cost}) + (\text{treatment cost})}{(\text{number of served households}) \times (\text{desludging period}) \times (12\,\text{months})}$$

The following tables show the calculation methods that we can use to obtain the basic cost of the scheduled desludging service. Information on cost items for collection and treatment can be obtained from a service provider, septage treatment plant manager or from the tariff regulations. Information on cost items for management can be obtained from organizations that have similar public service delivery functions, for example a water supply utility or company. The cost calculation examples below use numbers from previously calculated operating estimates.

(1) Obtain values of the parameters that have been calculated previously:

| *Parameter* | *Example* |
|---|---|
| • *Number of operation days (days/year)* | *250 days/year* |
| • *Desludging period (years)* | *3 years* |
| • *Number of served households (households)* | *48,000 households* |
| • *Number of served households per day (households/day)* | *64 households/day* |
| • *Number of served households in 1 operational cycle (households/cycle)* | *2 households/cycle* |
| • *Volume of septage that needs to be treated ($m^3$/day)* | *96 $m^3$/day* |

(2) Calculate transportation cost:

| *Parameter* | *Example of Calculation* |
|---|---|
| • *Desludging cost (USD per household)*[1] | *USD 10.00 per household* |
| • *Transportation distance of desludging truck (km/cycle)* | *40 km/cycle* |
| • *Fuel consumption (km/liter)* | *8 km/liter* |
| • *Fuel cost (USD per liter)* | *USD 0.5 per liter* |

*(Continued)*

[1]It is assumed that the amount is taken from the existing desludging service provider.

| *Parameter* | *Example of Calculation* |
|---|---|
| • *Number of operational cycles per day* | *(64 households/day)/(2 households/cycle) = 32 cycles/day* |
| • *Desludging cost item (USD per day)* | *(64 households/day) × (USD 1000 per household) = USD 640 per day* |
| • *Transportation cost item (USD per day)* | *(32 cycles/day) × (40 km/cycle) × (USD 0.5 per liter)/(8 km/liter) = USD 80 per day* |
| • *Collection cost (USD per day)* | *USD 640 per day + USD 80 per day = USD 720 per day* |
| • *Collection cost (USD per year)* | *(250 days/year) × (USD 720 per day) = USD 180,000 per year* |

(3) Calculate management cost:

| *Parameter* | *Example of Calculation* |
|---|---|
| • *Salary cost item (USD per year)* | *USD 90,000 per year* |
| • *Office overhead cost item (USD per year)* | *USD 120,000 per year* |
| • *Promotion cost item (USD per year)* | *USD 50,000 per year* |
| • *Depreciation cost item (USD per year)* | *USD 0 per year* |
| • *Management cost (USD per year)* | *USD 90,000 per year + USD 120,000 per year + USD 50,000 per year = USD 260,000 per year* |

(4) Calculate treatment cost:

| *Parameter* | *Example of Calculation* |
|---|---|
| • *Treatment fee (USD per $m^3$)* | *USD 3.00 per $m^3$* |
| • *Treatment cost (USD per day)* | *(96 $m^3$/hari) × (USD 3.00 per $m^3$) = USD 288 per day,* |
| • *Treatment cost (USD per year)* | *(250 days/year) × (USD 288 per day) = USD 72,000/year* |

(5) Calculate basic cost of scheduled desludging:

| *Parameter* | *Example of Calculation* |
|---|---|
| • *Total cost (USD per year)* | *USD 180,000 per year + USD 260,000 per year + USD 72,000 per year = USD 512,000 per year* |
| • *Basic cost (USD per month)* | *(USD 512,000 per year)/(48.000 households)/(12 months/year) = USD 0.9 per month* |

**Figure 3.1** With a good tariff, the accumulated revenue will be greater than the accumulated operating costs. The scheduled desludging scheme therefore can benefit the service provider.

By using a spreadsheet program, such as Microsoft Excel, the calculations to estimate operations can be combined with the calculations of basic cost. Different scenarios can be simulated quickly and easily with the spreadsheet program, any change in operating parameters will automatically adjust the basic cost.

## 3.4 STRIVE FOR PROFIT

In the end, all costs must be borne by the households in accordance with the principle of cost recovery. The rate to be charged to households must be at least the same amount as the basic fee. It is even better if the tariff is set higher than the base fee. Referring to the example above, we may propose the rate of USD 1.00 per month for the tariff to households which is slightly higher than the basic cost of USD 0.9 per month. So, the proposed tariff is not just to fully recover the cost but also to make the scheduled desludging operation bring some profit to the city or service provider (Figure 3.1).

The fact that a continuous stream of revenue is expected encourages decision makers in the city to accept the introduction of a scheduled desludging scheme in their city. It is important to convince the city government and other decision makers that the scheduled desludging scheme can be financially independent or even bring profit to the city and the service provider. They should understand that the scheduled desludging scheme will not place additional burdens on the local budget.

### Toolkit to simulate operation of the scheduled desludging scheme

A computer software has been created to help us simulate the operation plan of a scheduled desludging scheme and find out the implications for the financial aspects of the scheme. The official name of this software is the Septage Management Decision Support Tool (SMDST) but it is often referred to as the Septage Management Toolkit. Users can calculate the needs of the desludging fleet and septage treatment according to the projected number of served households. Different operation simulations can be performed by changing the values of the desludging technical parameters, such as septage volume, working time and distance to septage treatment plant.

This toolkit has five modules arranged in sequence: (1) flow and characteristics modules, (2) desludging modules, (3) processing modules, (4) management modules, and (5) financial modules. Based on the selected operating conditions, this toolkit will help the user to calculate the capital and operating costs of scheduled desludging. This will also help the user to calculate the basic cost and then the average tariff. The final output of the toolkit consisting of five modules is a financial plan, both in the form of a cash flow statement and an income statement. The results can be presented as pie charts.

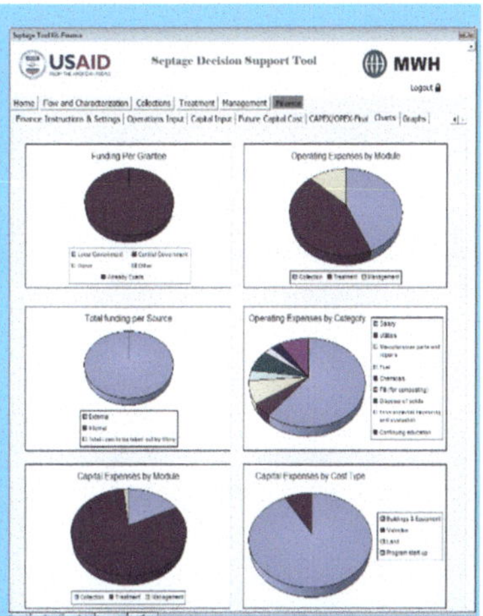


The user is guided to use the five modules, including entering the values of the operational parameters. If the user is not sure about the operational parameters, the toolkit provides a set of default values. The user can choose the septage treatment technology most appropriate for the specific conditions of a city, be it a mechanical system, non-mechanical system, or a combination of both.

*This toolkit was created by USAID by involving consultant Montgomery Watson Harza (MWH) as the developer. IUWASH is helping to improve this toolkit to make it more suitable for Indonesian cities. IUWASH is also fully involved in making the Indonesian version. This book can be downloaded from the IUWASH website.*

# Step 4
# Get the green light

*Once the team presents the initial concept of the scheduled desludging to the city government and other decision makers and convinces them that the scheduled desludging is needed and will bring benefits to the city, then the City government needs to give a green light for the preparation plan of the scheduled desludging scheme to move forward. Scheduled desludging is a new type of service for many cities. Whether it will be carried out or not depends on the decision of the city government and other city leaders.*

## 4.1 PERMISSION TO CONTINUE

Cities are not always required to have scheduled desludging schemes in their area. They have the option to introduce other sanitation services, including on-demand desludging and off-site or sewerage systems. The choice to have a scheduled desludging scheme may be based purely on the will of the city to provide better and safer sanitation service for households. Therefore, permission from the city government and other city leaders is very important for the process (see Figure 4.1). If the city government does not give permission, the team must immediately stop preparations to avoid wasting more resources. If the city government gives the green light for the plan, the team can move forward. If allowed, more city agents will be involved in the preparation and additional budget may be provided.

It is also very important for us to know the expectation and concern of the city government on the development and operation of the scheduled desludging scheme.

## 4.2 PRESENT PERSUASIVELY

The team should meet with the city government to present and discuss our plan to prepare the scheduled desludging scheme in the city. Prepare the presentation material, along with the handouts and other supporting materials. The presentation materials must be short, attractive and readable with limited words. Use consistent fonts, colors and backgrounds to maintain the clarity of the presentation


doi: 10.2166/9781789062281_0025

**Figure 4.1** The city government and/or responsible city leaders must believe that scheduled desludging services are very important for the area. The city leaders must have a general understanding of the benefits as well as the various risks that come with the preparation and implementation of the scheme.

(see Figure 4.2). We must use good quality images that reinforce and complement the message. Things that should be covered in our presentation material at least are:

- Existing condition of wastewater management in the city includes information about level of access to latrine, use of septic tanks, availability of offsite system as well as availability of desludging and septage treatment service.
- Environmental and public health conditions include information about the condition of groundwater and river water in urban areas, along with the level of disease generation associated with poor sanitation conditions.
- Driving factors for cities to have scheduled desludging includes laws and regulations requiring sanitation services, benefits of sanitation services, sanitation services in other cities, support available from other parties.
- Basic understanding of scheduled desludging; includes information about its definitions, differences with other services, nature of scheduled desludging services, aspects and benefits of scheduled desludging.
- Scheduled desludging scheme to be developed: includes scheduled desludging illustrative diagrams showing service coverage, scale of operations, infrastructure needs, institutional framework and the year scheduled desludging is expected to start.
- Financial aspects of scheduled desludging: includes basic tariff and estimated revenue and expenditure within a certain time frame.

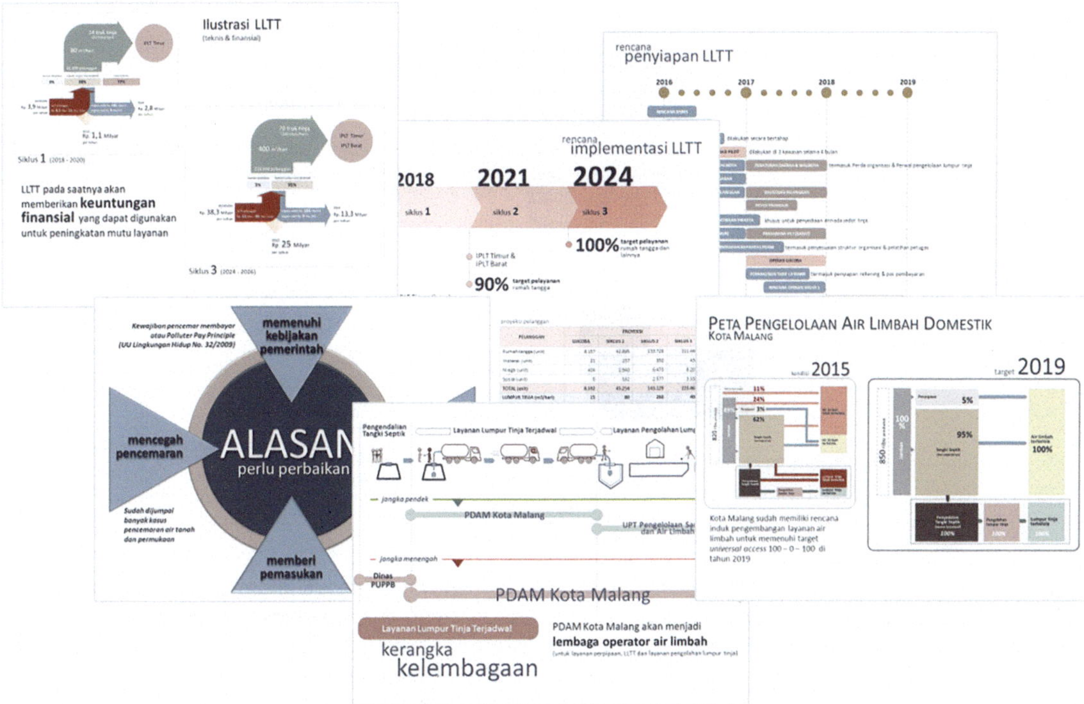


**Figure 4.2** Presentation material for the city government must be short and easy to understand, as well as an attractive layout design. The example above is the PowerPoint slides used by IUWASH program in Indonesia when presenting the initial concept of scheduled desludging scheme before the mayor of Solo (Central Java, Indonesia).

- Readiness of the city to prepare a scheduled desludging scheme: includes number of households with septic tanks, regulatory and institutional framework, availability of desludging trucks and septage treatment plant, other public services.
- Risk of scheduled desludging application: covers various technical, social, financial, and political risks that might occur if scheduled desludging is developed and implemented in a city.
- Workplan of scheduled desludging preparation: includes an outline of the work steps for the preparation and the duration of its implementation.

Convince the city government that the scheduled desludging is really needed in the city. Although there are risks in their development and application, the impacts of not having a scheduled desludging scheme in place will cause even greater losses, including environmental conditions, public health and welfare, financial and on the image of the city.

## 4.3 SHOW IT GRAPHICALLY

Use graphical diagrams to show the scheduled desludging concept. Try to include all information in one diagram (see Figure 4.3). Starting from the number of customers to the septage treatment capacity, as

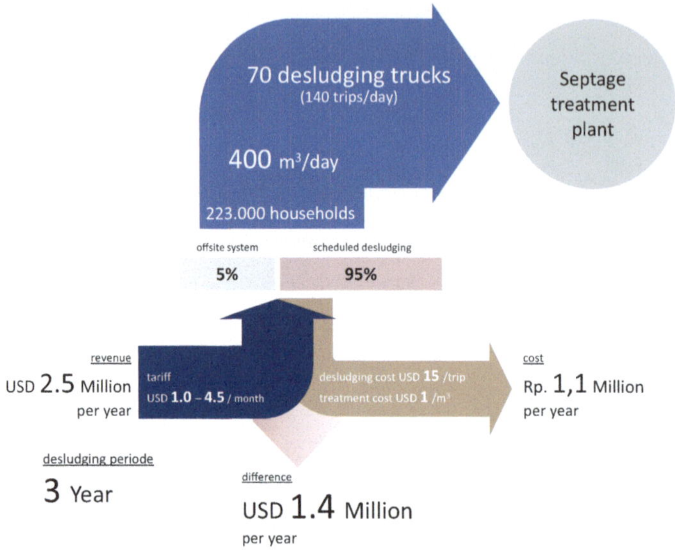


**Figure 4.3** The initial concept can be displayed in a diagram showing various information of the scheduled desludging scheme to be developed in a city. It contains information on the number of households to serve, desludging period, number of desludging trucks and septage treatment plant needed. In addition, this diagram also needs to include the basic tariff, operating cost and revenue projections.

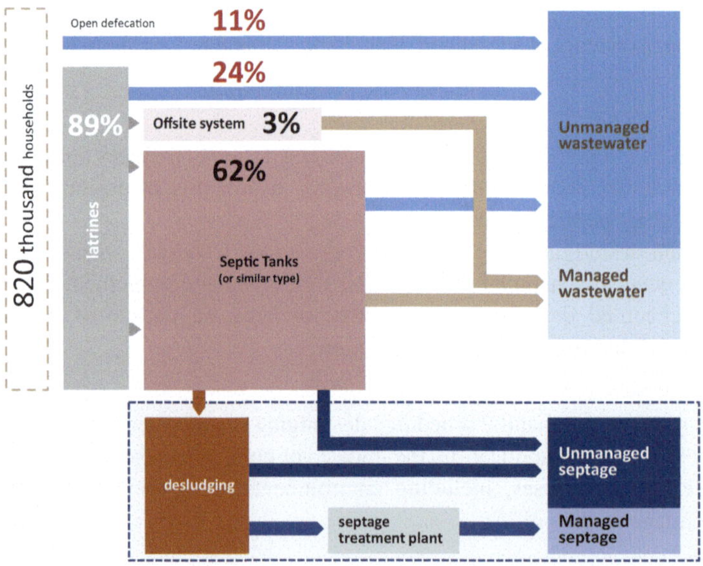


**Figure 4.4** The diagram shows the flow of wastewater produced by the entire population of the city. Some wastewater will be managed well, some will pollute the environment. Likewise, with the septage taken from septic tanks, some will go to the environment untreated. This flow diagram is developed by the World Bank and has been used in various cities in Indonesia and other countries.

well as financial information. Indicate the proportion of scheduled desludging components in the domestic wastewater management system.

In addition to the initial scheduled desludging scheme, provide another diagram that shows the overall wastewater management framework in a city (see Figure 4.4). This becomes more important in a city that has or will have an offsite system. In the wastewater management framework diagram, the context of the availability and contribution of scheduled desludging will be clearer.

## 4.4 RECORD THE DECISION

The city government's approval must be stated and recorded in writing. Particularly, if his or her permission is accompanied by promises to provide more technical, policy and financial supports. Prepare meeting minutes describing all the inputs and decisions made in the meeting. Send the meeting minutes to the attendees and other interested parties to ensure everybody will move forward from the same point.

# Step 5

# Assess targets

*Once the city government has agreed the scheduled desludging scheme can be prepared in the city, the team can enter the real planning process. Actual data must be used, specifically data on the number, condition and accessibility of septic tanks. Perform an extensive survey of most or all households and buildings in the city. Use a smartphone application to log the information and determine the targeted households and buildings to be served by scheduled desludging services.*

## 5.1 USE CURRENT INFORMATION

The team can use secondary information in making the initial concept of the scheduled desludging scheme (see **Step 3: Make initial concept**). However, in the advance planning of the scheduled desludging, it is strongly recommended to use actual information of the households (and other buildings) and their septic tanks. Scheduled desludging operation requires clear and accurate information of every service target by name and by address.

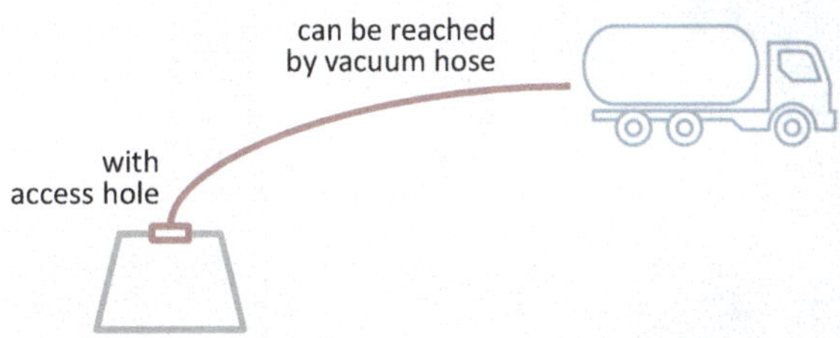


**Figure 5.1** A septic tank is considered viable for desludging if it has an access hole with a lid that can be opened and can be reached by the vacuum hose of desludging truck.


doi: 10.2166/9781789062281_0031

Information required in the planning of a scheduled desludging scheme includes:

- The number and distribution of buildings using septic tanks.
- The number and distribution of buildings using septic tanks that can be desludged (see Figure 5.1).
- Average septic tank volume, for each type of building.

All the information above should be provided as a total and as classified by types of buildings (residential or households, public offices, commercial buildings, and social buildings).

## 5.2 PERFORM SURVEY

The most effective way to identify potential scheduled desludging targets in a city is by conducting a survey of all households and buildings, or more commonly referred to as a census. This means that the team needs to go to and collect data directly from each building in the city. It certainly requires a lot of energy, time and funds since a city might have tens of thousands or even hundreds of thousands of buildings.

Alternatively, the team survey only some of the buildings which may represent the availability and condition of all septic tanks in the city. The number of buildings to be surveyed is determined by considering the time and funds available, of course without compromising statistical principle and validity. From the results of the survey, data can be interpolated to obtain information that describes the general condition of scheduled desludging targets in all areas of the city.

It is best to conduct a survey directly by visiting the buildings. In this way, an enumerator is able to interview the building owner or resident in person. The enumerator can explain the purpose of the question, record the respondent's non-verbal behavior and verify the answer visually. By visiting the location, the enumerator can obtain the GPS (global positioning system) coordinates of the septic tanks. The disadvantage is the team must have large funds and a longer period of time. There is also the possibility that the enumerator can influence the respondent's answers during the face-to-face interview (Figure 5.2).

**Figure 5.2** The best way is to perform face-to-face interviews with the owners or the residents of the buildings. If it is not possible, telephone interviews or mail questionnaires are also options.

Surveys can also be carried out indirectly, for example, through telephone interviews and mail questionnaires. Both of these ways are certainly more efficient in terms of energy, time, and funds. However, both have the same limitations, that is, the enumerator cannot clarify answers visually or capture non-verbal behavior of respondents. A telephone interview will get a higher response rate than a survey by mail.

## 5.3 COLLECT DATA

The team needs to get the following data when conducting a survey of the targeted buildings:

- Regarding the building: (a) availability of septic tank, (b) type of building (household, government, social, commercial), (c) accessibility of the building, (d) subscription to electricity and water supply services.
- Regarding the septic tank: (a) sources of wastewater that go to the septic tank, (b) septic tank location and accessibility, (c) size or dimensions, (d) tank material, (e) condition of the lid, (f) condition of the ventilation, (g) outlet system, (h) last desludging.

Use questionnaire to help the enumerator in collecting data during the survey or census (see Figure 5.3 and **Annex B** for the questionnaire).

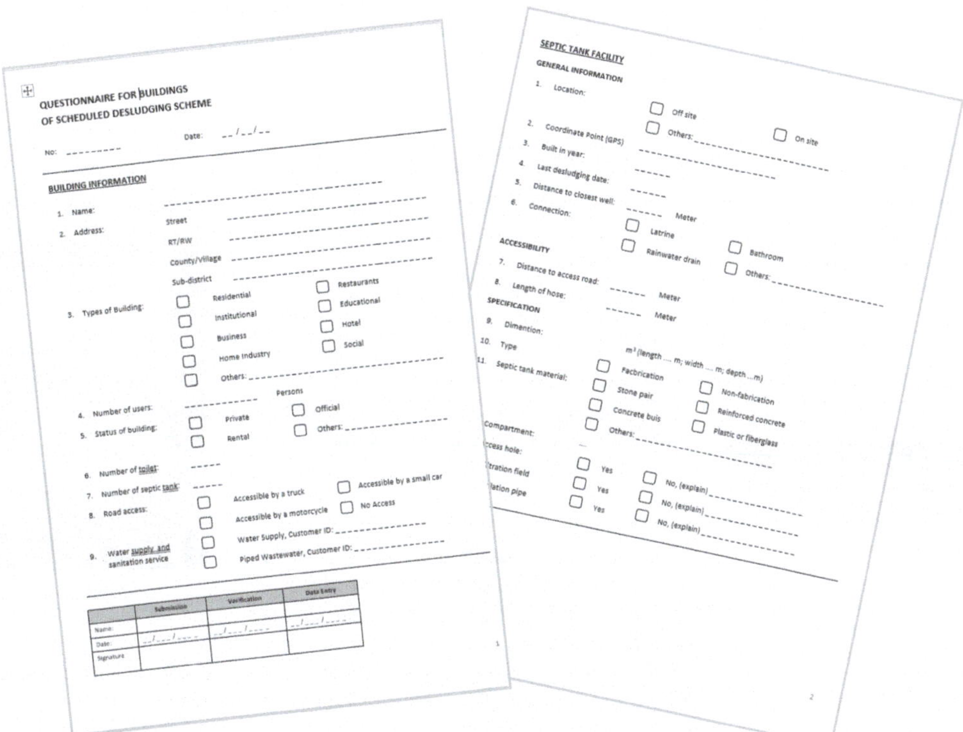

QUESTIONNAIRE FOR BUILDINGS
OF SCHEDULED DESLUDGING SCHEME

No: --------- Date: --/--/--

BUILDING INFORMATION

1. Name:
2. Address: Street / RT/RW / County/Village / Sub-district
3. Types of Building: Residential / Institutional / Business / Home Industry / Others / Restaurants / Educational / Hotel / Social
4. Number of users: ---------- Persons
5. Status of building: Private / Rental / Official / Others
6. Number of toilet:
7. Number of septic tank:
8. Road access: Accessible by a truck / Accessible by a small car / Accessible by a motorcycle / No Access
9. Water supply and sanitation service: Water Supply, Customer ID / Piped Wastewater, Customer ID

| | Submission | Verification | Data Entry |
|---|---|---|---|
| Name: | | | |
| Date: | --/--/---- | --/--/---- | --/--/---- |
| Signature | | | |

SEPTIC TANK FACILITY

GENERAL INFORMATION

1. Location: Off site / On site / Others
2. Coordinate Point (GPS)
3. Built in year:
4. Last desludging date:
5. Distance to closest well: ----- Meter
6. Connection: Latrine / Bathroom / Rainwater drain / Others

ACCESSIBILITY

7. Distance to access road: ----- Meter
8. Length of hose: ----- Meter

SPECIFICATION

9. Dimention: m³ (length ... m; width ... m; depth ... m)
10. Type
11. Septic tank material: Facbrication / Non-fabrication / Stone pair / Reinforced concrete / Concrete buis / Plastic or fiberglass / Others

Compartment:
Access hole: Yes / No, (explain)
Filtration field: Yes / No, (explain)
Ventilation pipe: Yes / No, (explain)

**Figure 5.3** The questionnaires will guide enumerators to obtain complete data in the survey. The structure of the questionnaire must be relevant, logical and concise. Use communicative terms, for example septic tanks rather than onsite units, toilet water rather than black water. Ask for the identity of the building owner or resident because the data will later be entered into the scheduled desludging operational database.

Digital applications are now available to assist enumerators in collecting data from respondents or potential customers (see the following box). Data obtained by the enumerator will be sent directly to the data management center via internet network.

## 5.4 ANALYZE FOR INFORMATION

The data collected needs to be analyzed in order to obtain the necessary information. start from the information on the percentage of buildings using a septic tank to the average septic tank volume and then move on to the technical parameter information. It is difficult to obtain a single number for an average septic tank volume, especially in a city where septic tank sizes vary widely. In such a condition, we can create average volume across several size classifications. For example, an average volume for small septic tanks, for medium-sized septic tanks and for large septic tanks. Use statistical programs, such as SPSS (Statistical Package for the Social Sciences) or Microsoft Excel, to accelerate the work. The results are displayed then presented in the form of tables, diagrams and narratives that are easy to understand.

There are thousands or even millions of data to be entered into a computer so there will always be a possibility of data reading or typing errors. There will always be questionable data due to misunderstanding in questioning and answering or the lack of thoroughness of the enumerator in recording the answer. Such data is better to be reconfirmed before entering into the statistic program.

### Android application for septic tank survey

Everyone with household-level survey experience may agree that the work of recording, entering, and processing data is very tiring. Especially if the survey is conducted on tens of thousands of respondents as in the survey of septic tanks for the planning of scheduled desludging in a city. In order to ease the task of the enumerator, while increasing the accuracy of data management, IUWASH has developed the Septic Tank Survey Application for Android-based smartphones.

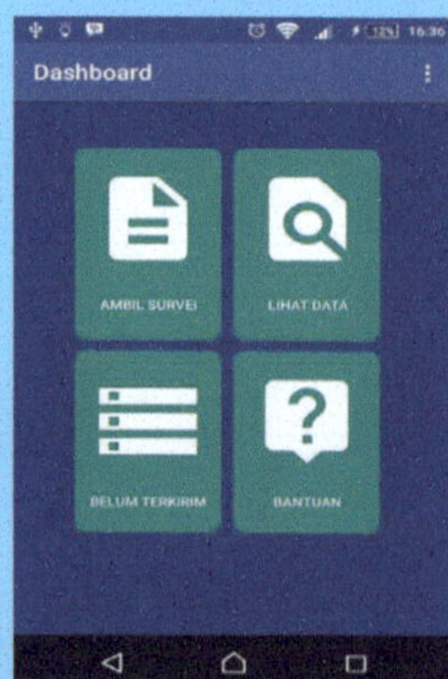


This application contains a series of questions relating to the characteristics of the septic tank and the building where the septic tank is located. Survey answers are directly filled in the application and then sent via the internet to the host computer. The Survey Portal application on the host computer will allow us to view survey results collectively.

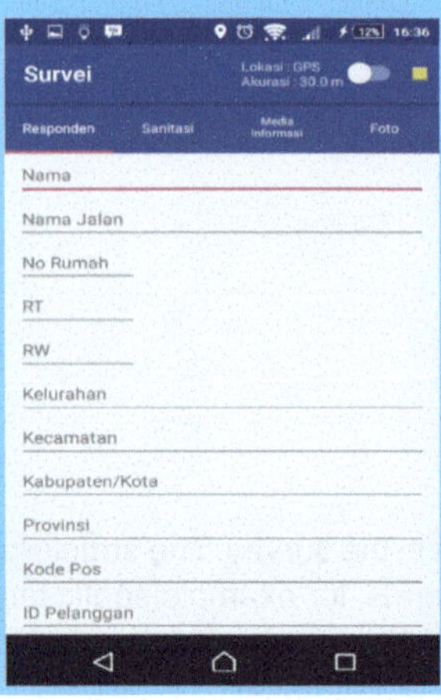


The use of smartphones makes the survey faster and enjoyable. Filling out the questionnaire, taking photos, and taking coordinates of the septic tank location can be done with one device. Immediately we can see the data in real-time, get a map overview of the situation of the area and the conditions of the septic tank on one smartphone.

In a flash, we can get the information needed to plan scheduled desludging operations, namely the number and distribution of buildings using septic tanks. The recording of the coordinates of the septic tank in the smartphone application allows us to find out the distribution of septic tanks in a city.

This smartphone app can also help us to monitor the progress of the survey implementation of each enumerator. If needed, the results of this application can be connected to the Management Information System (MIS) application that is used by scheduled desludging service provider. This integration is expected to simplify the process of receiving and managing scheduled desludging targets.

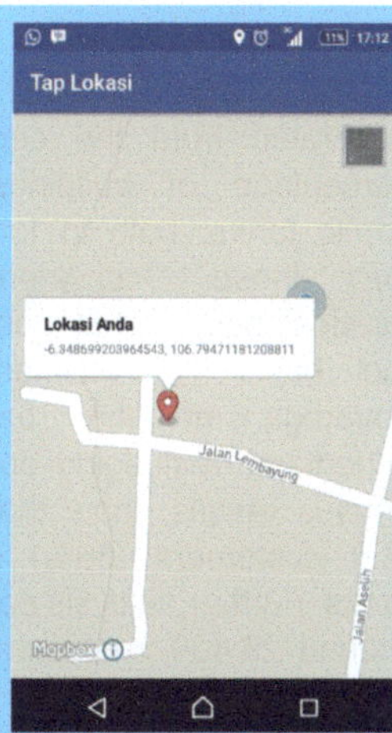

If the data has entered the statistic program in the computer, it can be analyzed to come to more actionable conclusions. For example, ‘the percentage of septic tanks in government buildings that have not been desludged in the last three years’ and ‘the percentage of septic tanks with impermeable walls’.

## Census of scheduled desludging potential customers in the city of Solo

The census for a scheduled desludging scheme is conducted for all water supply customers of the city of Solo who are not connected to the sewerage system. Around 48 thousand water supply customers were surveyed and the data that were successfully analyzed reached more than 41 thousand customers. The census was conducted at the end of 2014 and it took approximately 6 months to complete. Eighty enumerators were involved, half of whom were water supply meter readers.

The most important information generated from the census is the percentage of buildings that use septic tanks, both in total and only for those considered viable for desludging. That number will be the basis for estimating service coverage at the beginning of scheduled desludging operation. The census results show that nearly 35 500 customers (86%) have septic tanks. Further analysis concluded that around 26 400 customers (64%) are viable for desludging. This information will be more easily understood if it is presented in a pie diagram.

Before the census started, all enumerators took part in a training to understand the objectives, coverage, and procedures of the census. Topics in the training included (1) specification of septic tanks, (2) interview and observation techniques, (3) the use of questionnaires and (4) taking coordinates with GPS, specifically for the head of the enumerator group. During the census, the head of the enumerator group periodically cross-checks the 5% of all respondents who have been previously entered in the census.

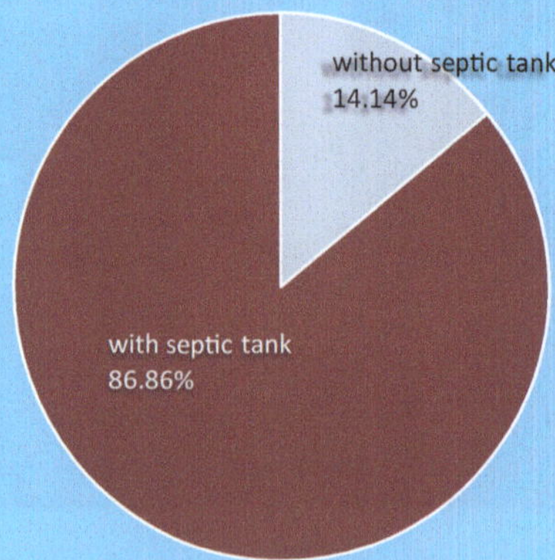


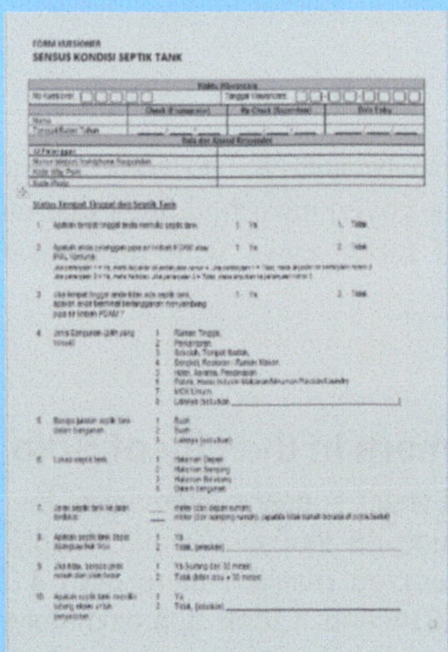

FORM KUESIONER
SENSUS KONDISI SEPTIK TANK

The survey also produced information to determine the function of the building, the condition and location of the septic tank, the volume of the septic tank and the coordinates of the location of the septic tank. It was concluded at the end that 66% of the septic tanks in the city of Surakarta had a volume above 3 meters. The results of the survey of prospective customers conclude with a workshop to discuss the results of the survey with various stakeholders in the city of Surakarta.

**Figure 5.4** Enumerators can carry out more than just collecting data from homeowners. They can also play a role in increasing the homeowner's knowledge of the correct use of septic tanks. The opportunity to meet directly with the owner or occupant of the house is a rare opportunity. Make the most of the opportunity.

## 5.5 USE FOR OTHER PURPOSES

The survey provides an opportunity to convey various information to prospective scheduled desludging customers, for example information about sanitation conditions in the area, proper septic tank specifications, scheduled desludging plans and others. Take advantage of this opportunity. Provide the enumerator with the information you want to convey. Create leaflets to help convey the information (Figure 5.4).

Take advantage of the survey data later as data for the scheduled desludging operation database. We already have information regarding the building and septic tank. Also use these data to develop scheduled desludging promotion strategies in the future. The government can also use these data to develop a septic tank control program or a program to subsidize the community in repairing or procuring septic tanks.

# Step 6

# Design operations

*By now, the team has actual information about the number of septic tanks in the city and the buildings that use them. The team also knows the capacity of the existing septage treatment plant which often becomes the limiting factor of a scheduled desludging operation. With all the information available, the team can develop the design of scheduled desludging operation in the city. Not only for the initial stage of operation, but also for the later stages. A combination of various modes of desludging and septage transportation can be considered in the design for service efficiency.*

## 6.1 CLASSIFY TARGETS

Scheduled desludging targets are classified according to the type of buildings that use the septic tanks, for example residential households, government or public offices, commercial buildings and social buildings (see Figure 6.1). Classification of scheduled desludging targets can also follow the classification of customers that already applies to other services. If the scheduled desludging is managed by the same institution that manages water supply, the classification of the scheduled desludging targets can follow the existing water supply customer classification.

Some cities have communal wastewater services that use shared septic tanks or other types of anaerobic treatment units that also require a regular septage desludging (see Figure 6.2). A separate customer categorization is often needed for group of buildings with such communal service. The registered target might be the community group that manage the service.

Classification of the targets is necessary to differentiate the services provided to each group and the service charge. For example, differentiation of desludging periods, maximum desludging volumes and customer rates. If there will be no distinction, no target classification is needed.


doi: 10.2166/9781789062281_0039

**Figure 6.1** The targets of scheduled desludging can be classified according to the use and function of the buildings. The classification can further be specified according to the size of the building and the location of the building. It can also be specified according to the condition of the area where the building is located, for example the width of the road and the status of the area.

## 6.2 DIVIDE SERVICE ZONES

The division of service zones will make it easier for the service provider to manage their targets and to reduce the movement of their desludging fleet. Service zones can be divided as follows:

- Administrative area: where the boundaries of a service zone follow the boundaries of one subdistrict or several adjacent villages.
- Existing service zones: for example, following water supply service zones in cities that will appoint the water supply service provider as scheduled desludging service provider.
- Distance radius to septage treatment plant: where a service zone is formed from areas within a 5 km radius of the septage treatment plant.

It is important to remember that service zoning is only needed if scheduled desludging scheme has either a very large area, a large number of customers or more than one septage treatment plant.

## 6.3 CONFIRM DESLUDGING PERIOD

Theoretically, the scheduled desludging period can be determined if we know (a) the average volume of septic tanks in the city, (b) the annual amount of sludge generation per person, and (c) the average number of residents of a building. In practice, it is impossible to get a correct number of desludging period due to the high variations of septic tank volume and number of residents. A desludging period of 2–5 years is common practice in cities throughout the world. Indonesian septic tank standard requires the

**Figure 6.2** Community-based wastewater system uses a treatment unit that also require periodic sludge removal services. If agreed, they can also be the target of the scheduled desludging service.

sludge compartment of a septic tank to be designed for a 3-year holding capacity (see Figure 6.3). Accordingly, many Indonesian cities plan to use the mandatory desludging period of 3 years.

The desludging period affects the volume of septage to be treated. The longer the desludging period, the less the daily amount of septage to be treated. The same effect also for the number of the desludging trucks required. The longer the desludging period, the less the number of desludging units needed. It is also possible that each building classification will have a different desludging period. For example, 3 years for households, 2 years for public offices and 1 year for commercial buildings.

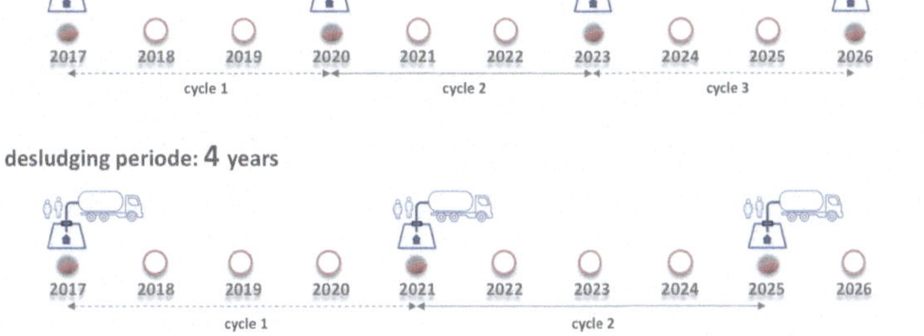


**Figure 6.3** Illustration of scheduled desludging period of 2, 3, and 4 years. The shorter the desludging period, the higher the desludging frequency is carried out and the more desludging units are needed.

## 6.4 SPECIFY LEVEL OF SERVICE

It is difficult to ensure periodic desludging services will be available for all buildings in the city since the initial cycle. The 100% level of service may only be achieved in the third or fourth cycle along with the readiness of the infrastructure and the service provider. Service level of scheduled desludging must be determined realistically with thorough considerations on:

- **Regulation**: If there is no mandatory requirement for periodical desludging, it is difficult for a city to set high service level.
- **Septic tank condition**: The fewer the number of viable septic tanks, it is difficult for a city to set high service level.
- **Septage treatment**: The capacity of septage treatment plant might be the determining factor of the city in setting the level of service, especially in the initial cycle.
- **Wastewater development policy**: If available, we need to refer to the existing and future coverage of sewerage system in the city before we can set the service level of scheduled desludging.
- **Capacity of service provider**: The higher the service level, the higher the demand for the service provider capacity. Not only concerning the staffing but also the availability of management system and facilities.
- **Financial goals**: Scheduled desludging can be planned for cost recovery but also can be planned for being subsidized. Level of service must provide revenues and expenses that are consistent with the financial plan.

The planned level of service determines the scale of scheduled desludging operation. The higher the service level plans, the higher the frequency of the desludging and transportation operation. The financial aspects of scheduled desludging follow the same pattern; the higher the level of service will be, the greater the financial transaction within the scheduled desludging scheme. In short, there are technical and financial implications to be considered when determining the level of service of the future scheduled desludging operation. Determination of service levels often requires iteration to ensure a service level target does not exceed septage treatment capacity or other constraints.

The team can plan the level of service for each building classification. Distinguishing the service level might be necessary, especially at the beginning of scheduled desludging implementation. When households are not considered ready, it is better for a municipality to prioritize scheduled desludging for public offices or other buildings (see the following table). Thus, the level of service for public buildings will be higher than for other building classification. In the next cycles, service level for households and commercial buildings will be increased.

Service targets for each customer classification

| Classification of Buildings | Targeted Level of Service (% buildings) | | |
|---|---|---|---|
| | Cycle 1 | Cycle 2 | Cycle 3 |
| Households | 10 | 60 | 90 |
| Public buildings | 80 | 85 | 90 |
| Commercial buildings | 10 | 80 | 90 |
| Social buildings | 0 | 60 | 90 |

## 6.5 DETERMINE MODES OF OPERATION

### 6.5.1 Three desludging modes

There are three modes of desludging that can be applied in a scheduled desludging scheme (see Figure 6.4): (a) overall desludging, (b) proportional desludging and (c) fixed volume desludging. The selection of a desludging mode will affect the number of trucks needed and the number of transportation trips to the septage treatment plant.

- In **an overall desludging mode**, the service provider will remove all septage from the septic tank. This desludging mode will allow the septic tank to regain its maximum storage volume until the next desludging event. Given the high variation of septic tank volume, it is difficult for the service provider to make the trucks work efficiently.
- In **a proportional desludging mode**, the service provider will remove septage with a fixed volumetric proportion of the septic tank. For example, with a 60% proportional desludging, the desludging truck will extract 1.2 $m^3$ of septage from the 2 $m^3$ septic tank or remove 1.8 $m^3$ of septage from a 3 $m^3$ septic tank. Technically, the proportional desludging mode has the best basis for consideration. However, the application of this mode will face the same inefficiencies as the overall desludging mode. In addition, it is difficult for desludging worker to ensure that the septage has been removed in proportion to the volume of the septic tank.
- In **a fixed volume mode**, the service provider will remove septage with a constant volume from the entire septic tank. For example, with a fixed desludging of 1.5 $m^3$, the desludging truck will remove the same volume of 1.5 $m^3$ of septage from the 2 $m^3$ septic tank or from the 3 $m^3$ septic tank. The operations of the desludging fleet can be more efficient with this fixed desludging mode. The disadvantage is this mode does not return the volume of the septic tank to its maximum holding capacity.

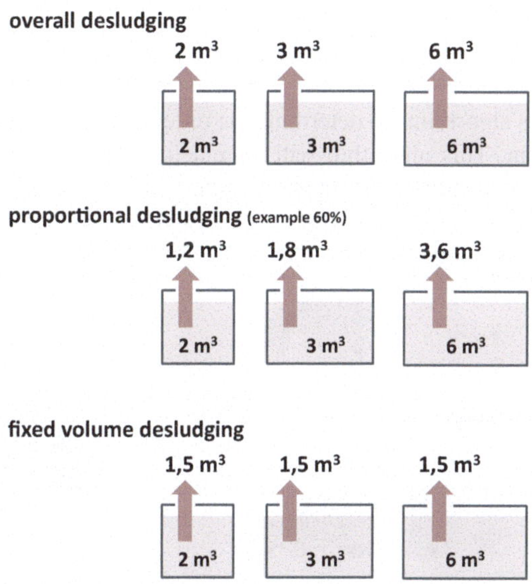


**Figure 6.4** There are at least three desludging modes that can be selected for scheduled desludging scheme. Each has its own advantages and disadvantages. In the initial stages, it helps to use the simplest mode, that is, the fixed volume mode.

Use information of the average septic tank volume obtained from the survey, before selecting the proportional volume mode or the fixed volume mode. The technical and financial risks of each selection can be minimized by choosing the proper tank dimension of the desludging truck. Please be aware that the households need to be informed of the reason a particular desludging mode is selected.

### 6.5.2 Two transportation modes

There are two transportation modes that a city can apply to bring the septage to the specified treatment facility (see Figure 6.5): (a) direct transportation and (b) collective transportation. The selection of transportation mode will affect the number of desludging trucks needed and the number of trips required to transport the septage to treatment facility.

**The direct transportation mode** is applied if we want to transport the septage by the desludging unit that just previously carried out a septic tank desludging operation. This transportation mode is suitable for service zones that are close to the septage treatment facility. For service areas that are far from the treatment facility, for example those over 20 kilometers away, the application of **the collective transportation mode** is worth considering. Desludging units will collect septage in a transfer station, or temporary sludge storage (TSS), before a large sludge truck brings the septage to the treatment facility. A mobile transfer station has the advantage that it can be used in several locations while a fixed mobile transfer station ensures better reliability. The application of the collective transportation mode will reduce the number of fleets moving towards the septage treatment facility. Traffic density at STP is automatically reduced.

Septage can also be transported to septage treatment facility through the available sewer pipelines. However, this practice is allowed only if there is a special solid–liquid separation unit in the septage receiving point. The unit will reduce significantly the solid content in the septage before it enters the piping system. This mode is certainly only applicable in cities with sewerage systems that still have idle capacity.

### 6.5.3 Three scheduling algorithms

Scheduled desludging requires algorithms to determine the order in which septic tank desludging is carried out over a specified time span. This algorithm will later help the service provider to determine which buildings the desludging will be conducted at on an operation day. Some scheduling algorithms are:

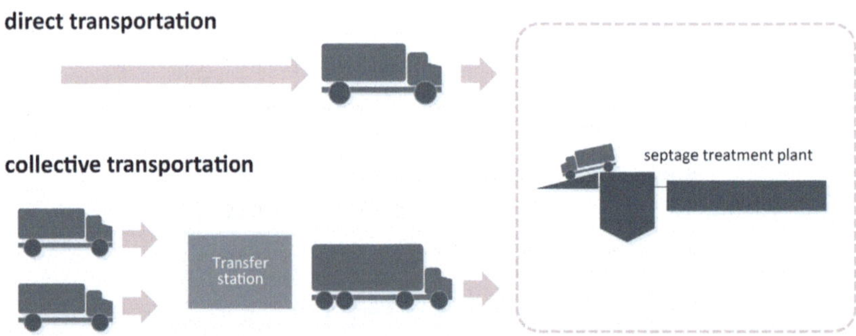


**Figure 6.5** Scheduled desludging scheme can apply direct and/or collective transportation modes. The selection of the mode is influenced by the distance to septage treatment plant, number of trucks available and the road accessibility.

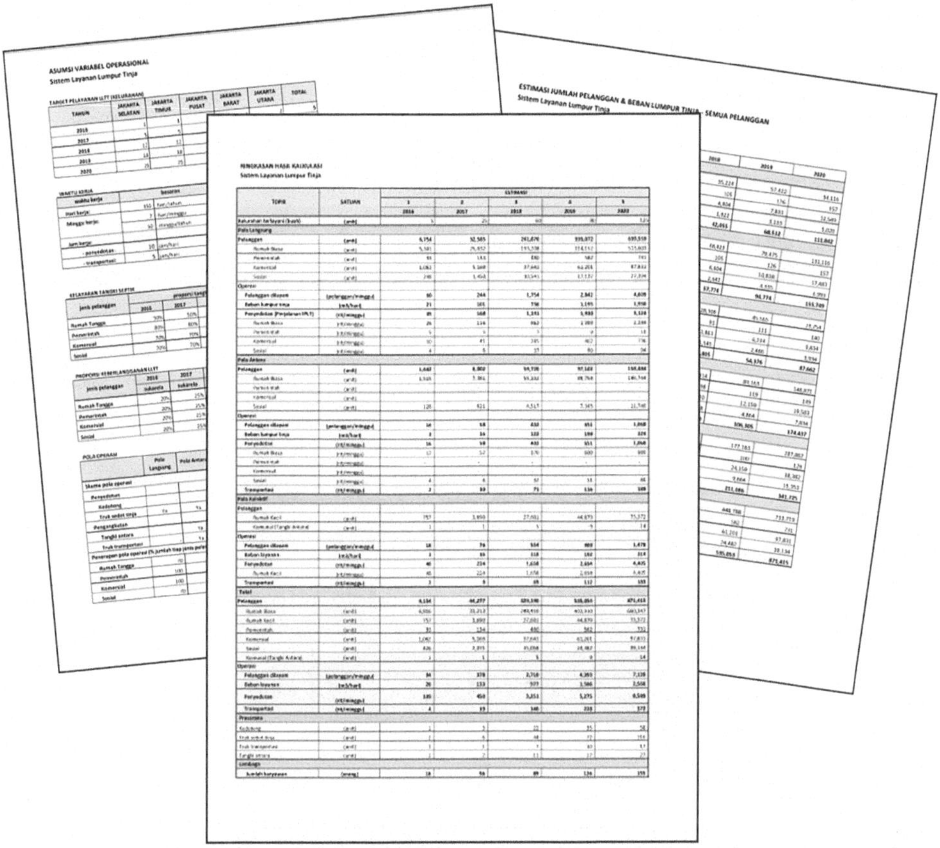

**Figure 6.6** Scheduled desludging simulation needs to use the help of spreadsheet software such as Microsoft Excel since the calculations performed can be very large and complex.

- **By area**: Desludging is carried out for all septic tanks in one area before moving to another area. The determination of the work area should follow the division of territory in a city as well as take into account the division of service zone.
- **By building classification**: Desludging is carried out for all septic tanks in buildings belonging to a particular classification before continuing to another classification. For example, septic tank desludging is scheduled for public buildings in semester 1, commercial buildings in semester 2, households in semesters 3 and 4.
- **By distance**: Desludging is carried out for all septic tanks located within a certain distance (radius) from the septage treatment plant before moving to the next radius. For example, septic tank desludging is scheduled for buildings within the radius of 0–3 km in semester 1, radius of 3–4 km in semester 2, then radius 4–5 km in semester 3.

The service provider can combine these two or three modes to create more efficient scheduling operation that are within their capabilities. For example, combining area-based scheduling mode with building classification scheduling mode.

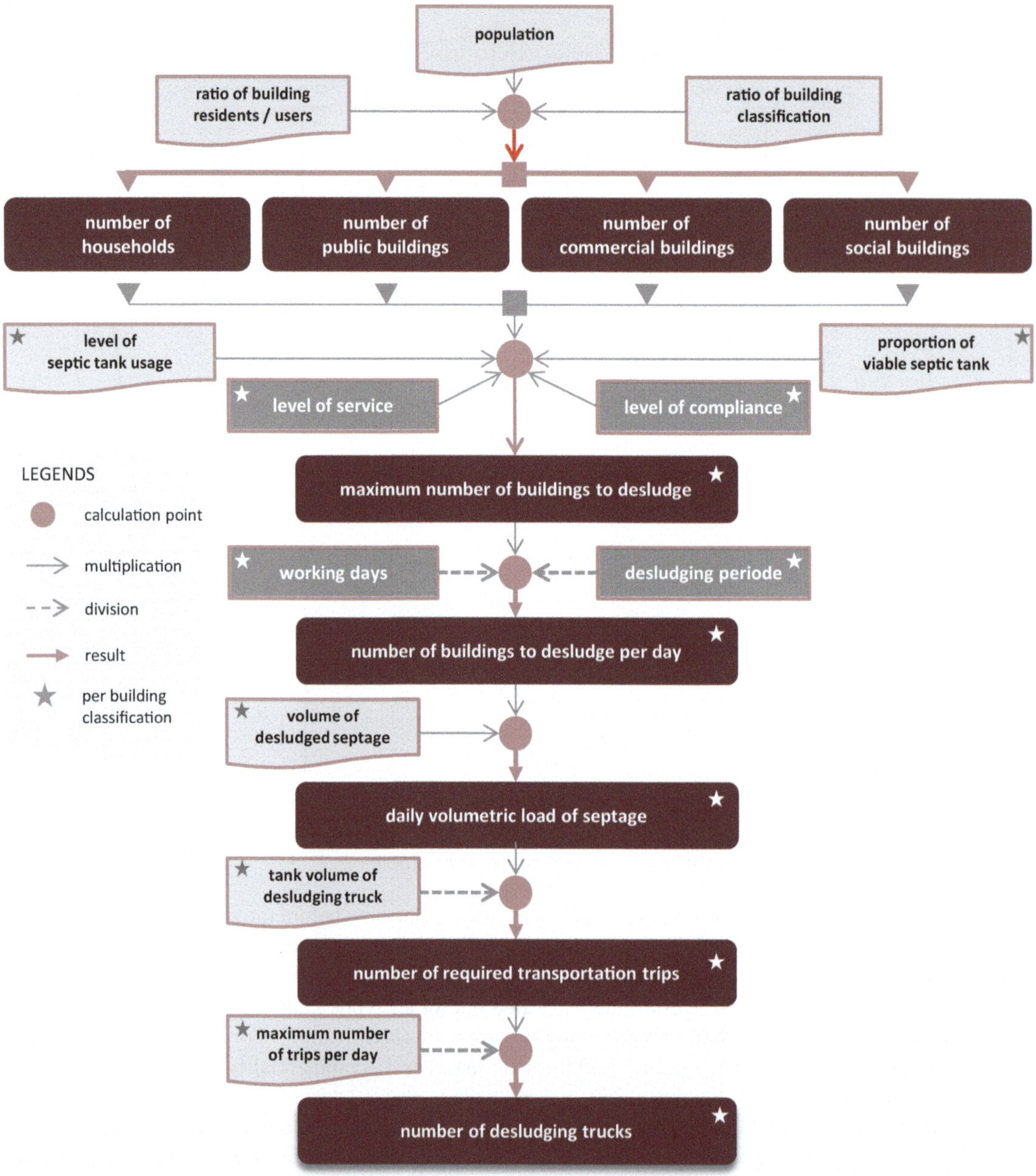


**Figure 6.7** Simulation flow diagram for scheduled desludging operation. Calculations are made for each building classification that might have its own targeted service level. The calculations in the simulation actually are similar to simple calculations in the making of the initial scheduled desludging concept. The difference is that operating simulations use more diverse operating modes and more actual technical parameter values.

## 6.6 SIMULATE OPERATION

The scheduled desludging operation should be simulated to estimate the desludging and transportation workload in accordance with the targeted level of service for an operating cycle. The simulation will also determine the requirements of desludging units (trucks or pick-up) and septage treatment facilities. There are several operating parameters with known values before we can simulate the operation, namely (a) the ratio of building residents, (b) proportion of building types, (c) proportion of septic tank usage, and (d) average volume of septic tanks. We can use the number of buildings to serve from the results of the survey or census. As for the next cycles, the number of buildings to serve must factor in the population growth and the change of service levels.

After determining the simulation time span, for example 2020–2030 or cycles 1–4, project the number of buildings for each operation cycle. Next, just follow these steps (see Figures 7.6 and 7.7):

- Calculate the maximum number of buildings to desludge (those using viable septic tank) for each classification with this general formula:

  $$\text{Maximum number of buildings to desludge} = (\text{number of buildings}) \times (\text{proportion of septic tank usage}) \times (\text{proportion of viable septic tank})$$

- Calculate the number of buildings to desludge. Use the targeted level of service with assumed compliance level with this general formula:

  $$\text{Number of buildings to desludge} = (\text{maximum number of buildings to serve}) \times (\text{level of service}) \times (\text{assumed compliance level})$$

- Calculate the number of buildings to desludge in one day. Use the number of working days per year and the confirmed desludging period in this general formula:

  $$\text{Number of buildings to desludge per day} = (\text{number of buildings to desludge}) / [(\text{desludging period}) \times (\text{working days in a year})]$$

- Calculate the volume of septage (which later required treatment). Use the number of buildings to desludge per day and the average volume of desludged septage in this formula:

  $$\text{Daily septage load} = (\text{number of buildings to desludge per day}) \times (\text{desludging volume})$$

- Calculate the number of transportation trips to the septage treatment plant. Use the daily septage load and the volume of desludging truck tank in this formula:

  $$\text{Number of required transportation trips} = (\text{daily septage load}) / (\text{tank volume of desludging truck})$$

- Calculate the number of desludging trucks required. Use the number of septage transportation trips and the maximum number of trips per day in this formula:

  $$\text{Number of desludging trucks} = (\text{Number of required transportation trips}) / (\text{Maximum number of trips per day})$$

**Annex H** contains an example of a series of spreadsheets made to plan the operation.

# Step 7

# Improve institutions

*Scheduled desludging requires an institution with technical and managerial capacity to deliver the service. An existing institution should be considered to be appointed as the main service provider besides the existing water utility. The organizational structure should match the needs of the scheduled desludging operation, as well as the amount and competence of the personnel. There needs to be an institution with the role to regulate the scheduled desludging operation.*

## 7.1 ASSIGNING EXISTING INSTITUTION

Since scheduled desludging is to be required by law, it is likely there will be a large number of septic tanks to be served in the city. Medium to large cities in Indonesia, such as Solo, Balikpapan, Medan, Bandung, Surabaya, and Makassar, require their scheduled desludging service to cover hundreds of thousands or even millions of septic tanks. Correspondingly, a scheduled desludging service must be managed by an institution that has the capacity and experience to serve a large number of customers. Not only to do the desludging operation in thousands of buildings, but just as important is to manage the service billing.

Most cities have an institution that provides on-demand desludging service or operation of the septage treatment plant. There are also cities with an institution to provide wastewater service using a sewerage system. As long as the institution has the capacity and experience to serve thousands of buildings, it can be considered for appointment as the main service provider for the scheduled desludging scheme. The main service provider is the one with authority to coordinate other services providers, particularly those from private sectors (see **Step 9: Involve partners**). By appointing an existing institution, the operation of scheduled desludging can share the available resources and facilities. Municipalities no longer need to provide additional resources and facilities as much as those for new institutions. However, the capacity of the existing wastewater institution can be the limiting factor to the level of service. This may cause the scheduled desludging operation to be implemented gradually. In the first years, scheduled desludging will only target a few buildings while the capacity of the service provider is strengthened.


doi: 10.2166/9781789062281_0049

## 7.2 CONSIDER WATER UTILITY

The only municipal institution with capacity to serve tens of thousands of buildings may be the water utility. As an institution that has been around for years or even decades, the water utility may have offices, administration, financial systems, staff and facilities to support a scheduled desludging operation (see Figure 7.1). By appointing the water utility, the scheduled desludging scheme can piggyback on the existing billing system; it can be combined or merged with the water bill. In this case, a building would only receive one monthly bill to cover tariffs or fees for water and scheduled desludging services. Building owners will inevitably pay for scheduled desludging service if they want to maintain their subscription to water service.

### Assigning the scheduled desludging to a wastewater utility in the city of Makassar (Indonesia)

The City of Makassar (South Sulawesi, Indonesia) launched a scheduled desludging scheme in August 2013. The scheduled desludging operation is assigned by the city government to an existing institution, the Technical Implementation Unit of the Wastewater Management Office (UPTD PAL, or *Unit Pelaksana Teknis Dinas – Pengelolaan Air Limbah*). UPTD PAL, with a total of 22 personnel, is led by a head who oversees four divisions which are responsible respectively for sewerage system, septage treatment, community-based system, and monitoring. The scheduled desludging operation is put under the septage treatment section.

The UPTD PAL was involved in the planning and preparation of the scheduled desludging scheme since it was introduced in Makassar in 2012 by IUWASH program. Their role included preparing a scheduled desludging concept and advocating it to the city government and other city leaders. In August 2013, UPTD PAL conducted a survey to 300 selected households. Almost all UPTD PAL staff were involved in the preparation processes to inform households, define service areas, conduct the surveys, designing operations, develop database with Geographic Information System (GIS) data. On the other hand, IUWASH program provided technical assistance to prepare SOPs, tariff modeling, improve customer database, and prepare the business plan.

Considering its capacity, UPTD PAL proposed to initiate the scheduled desludging operation only targeting 300 households in a housing estate. Besides conducting the desludging operation, UPTD PAL is responsible for collecting scheduled desludging service fees. They plan to add additional fleets, facilities and staff to support their future operation.

Not all water utilities can effectively manage a scheduled desludging operation. If the utility has less than 50% water service coverage, it is likely to be limited in access and capacity to incorporate the scheduled desludging scheme. Likewise, if the utility has more than 60% water service coverage, it should be considered.

**Figure 7.1** Many water utilities already use a management information system (MIS) where data of the subscribers are stored. Some MIS are integrated with geographic information systems (GIS) that show the spatial locations of the buildings. The scheduled desludging operation can benefit from the existing MIS and GIS.

The weaknesses of a water utility in managing scheduled desludging are:

- A water utility may prioritize their resources for their main and original assignment – the water supply service.
- A water utility may not have the capacity to conduct scheduled desludging operations.
- A water utility by regulation may be prevented from managing and delivering other services.

If a water utility and other existing institutions are considered not ready to manage the scheduled desludging operation, a new institution can be formed. More energy and time are needed, but there are several advantages to establishing a new institution. With new institutions, the organization can be designed in a way that is suitable to a scheduled desludging operation; The organization is not limited to methods that are already owned or standardized in an existing institution. The formation of a new institution will also create momentum for improving sanitation services more thoroughly. In addition to desludging services, a new institution should also be assigned to operate septage treatment and a sewerage system, if applicable. Such a multi-service institution offers more effective coordination between onsite and offsite systems. Utilization of resources will be more efficient and effective.

Assigning a scheduled desludging operation to either a water utility or wastewater utility will have advantages and disadvantages. Unlike a water utility that can combine scheduled desludging bill with water service bill, a wastewater utility does not have enough power to force building owners to pay a scheduled desludging bill on time. On the other hand, the water utility does not necessarily have sufficient technical capacity to carry out the desludging operation. Therefore, collaboration of these utilities may offer a more effective and efficient scheduled desludging operation. For instance, the wastewater utility may conduct the technical operation while the water utility provides supports for customer and billing management.

## 7.3 SEPARATE FUNCTION TO REGULATE

The achievement of goals and sustainability of scheduled desludging scheme is not only determined by the service provider. It is important in a good scheduled desludging or any wastewater institutional setup that the service provider is supported by other institutions with roles to:

- determine implementation policies,
- control the use of septic tanks,
- set target of scheduled desludging service coverage,
- enforce the obligation to desludge septic tanks and
- supervise the operational performance.

It is possible that one institution will carry out more than one task. For example, the agency of public works is given the roles to control the use of septic tanks and to supervise the performance of scheduled desludging operation.

The service provider must not carry out any of the roles above. A separate institution must be appointed to determine the policies or supervise scheduled desludging operational performance. By separating roles of regulator and service provider, conflicts of interest will be prevented. This principle must be maintained in the proposed scheduled desludging institutional framework.

Tasks within a service provider of scheduled desludging.

| Tasks<br>General | Tasks<br>Finance |
|---|---|
| 1. General administrative | 1. Budget and expenditure planning |
| 2. Secretarial duties | 2. Bookkeeping |
| 3. Procuring equipment and materials | 3. Billing management |
| 4. Office management | 4. Receiving payments |
| 5. Controlling assets | 5. Cost verification |
| 6. Managing work partners | 6. Payment or gratuity assignment |
| 7. Public relations | 7. Financial supervision and audit |
| 8. Legal affairs | 8. Evaluating financial performance |
| 9. Staffing | |
| **Customer** | **Technical** |
| 1. Marketing & promotions | 1. Scheduling operation |
| 2. Customer acceptance | 2. MIS operationalization |
| 3. Customer verification & mapping | 3. Carrying out operations |
| 4. Managing customer data | 4. Monitoring operations |
| 5. Customer relations | 5. Fleet maintenance |
| 6. Handling complaints | 6. Controlling environmental impacts |
| 7. Enforcement | 7. Occupational health and safety |

## 7.4 THE TASKS DEFINE THE STRUCTURE

The organizational structure of a scheduled desludging service provider must be arranged to fit the tasks. Not only technical tasks, but also administrative, customer and financial management tasks (see the following table). A wastewater utility that offers a scheduled desludging service will have a different organizational structure from a utility that offers several wastewater services. The organizational structure also differs between utilities with large and small operations, or between utilities with authority to manage finances and not. In the end, all the necessary tasks should be included in the organizational departments or units of a service provider.

The number of staff in a scheduled desludging service provider will be adjusted according to the scale of the operation. Not always directly proportional, but generally the more the number of targeted buildings and the wider the service area, the greater the number of staff is required.

# Step 8
# Set regulations

*The goal is for the septage to be pumped regularly from every septic tank. Only a designated and licensed service provider can perform the scheduled desludging operation. All buildings that receive periodic desludging must pay the tariff set by the municipality. Treatment of the septage must be conducted until the effluent meets a specified environmental standard. All of the above will only happen if there are regulations. Promotion and compliance strategies must be prepared for the regulations.*

## 8.1 MUST BE MANDATORY

Scheduled desludging is a mandatory act. Regardless of how empty or full its content, a septic tank must be desludged periodically at the specified time. This mandatory requirement is the essence of scheduled desludging which distinguishes it from the regular desludging service that households and building owners take for practical and economic reasons. Enforceable regulation is needed in a city before they can initiate a scheduled desludging scheme in the area. The regulation also needs to mention that the scheduled desludging operation will be carried out only by a designated service provider appointed by the municipality.

A city however should not make regulation specifically only for scheduled desludging. It is recommended that all scheduled desludging provisions should be parts of a more comprehensive regulation on septage or fecal sludge management. In other parts, the regulation should contain provisions related to septic tanks, septage transportation, septage treatment, and reuse of treated sludge (Figure 8.1).


doi: 10.2166/9781789062281_0055

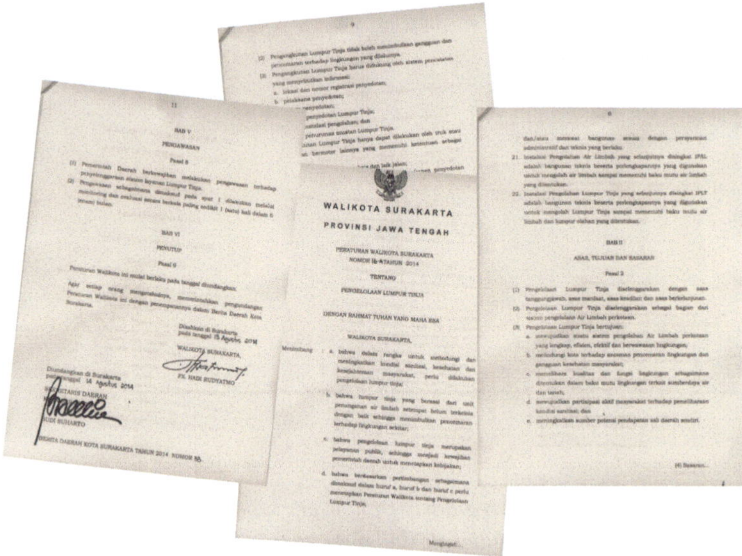


**Figure 8.1** The Mayor Regulation of Surakarta No. 16A of 2014 on Septage Management clearly puts periodical desludging as a mandatory measure for every septic tank user. It requires each septic tank to be desludged every 3 years. The regulation also contains provisions on institutional, technical, and financial aspects.

## 8.2 MANY THINGS TO COVER

The septage management regulation must contain a number of provisions which either directly or indirectly support the development and operation sustainability of scheduled desludging in the city (see Figure 8.2). These provisions include (a) prerequisites of a good septic tank, (b) institutional setup of septage management, (c) private sector partnerships in septage management, (d) implementation of on-demand desludging service, (e) billing and payment mechanism, (f) septage treatment effluent quality standard and (g) treated sludge quality standard. See **Annex C** for the example of a regulation on septage management.

The regulatory framework for septage management must also contain provisions regarding the determination of desludging tariff, both for scheduled and on-demand desludging. To anticipate frequent tariff changes, cities must make specific regulation regarding scheduled desludging tariffs. The same applies to the level of septage treatment. As such, any changes in scheduled desludging and septage treatment tariffs will not require that the whole regulation on septage management be changed.

## 8.3 REGULATIONS ARE HIERARCHICAL

Provisions for domestic wastewater and septage management must be placed in a series of hierarchical regulations (see Figure 8.3). The definition, obligations and guidance relating to septage management should be placed in the regulation of a mayor or district head or governor. Changes to the provisions of the regulations at this level do not require parliamentary approval. It will be easy and quick for the municipality to adjust the septage management provisions in the future. As shown in the previous

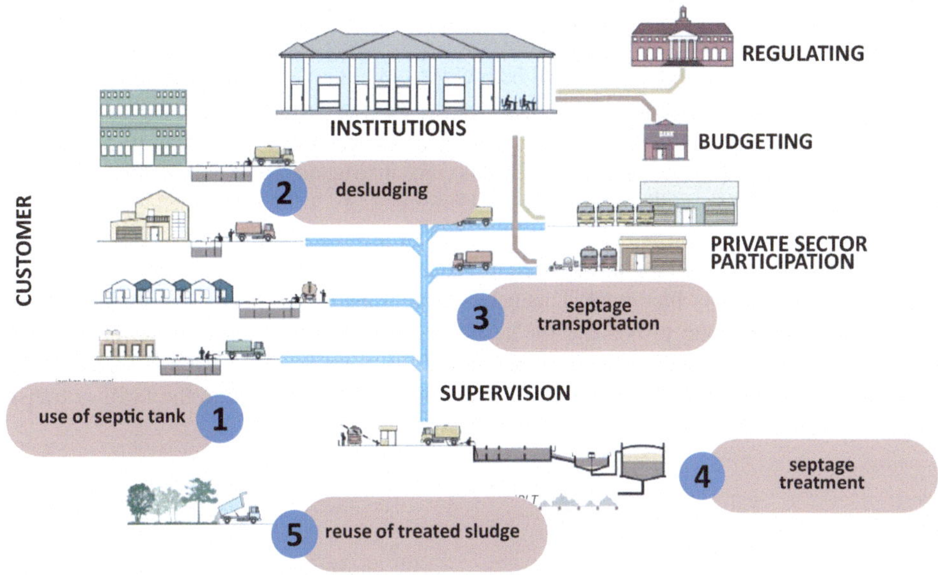


**Figure 8.2** The regulation on septage management must cover the issues shown above. Obligation to regularly desludge each septic tank should be included in the regulation to allow a scheduled desludging scheme to be implemented in a city.

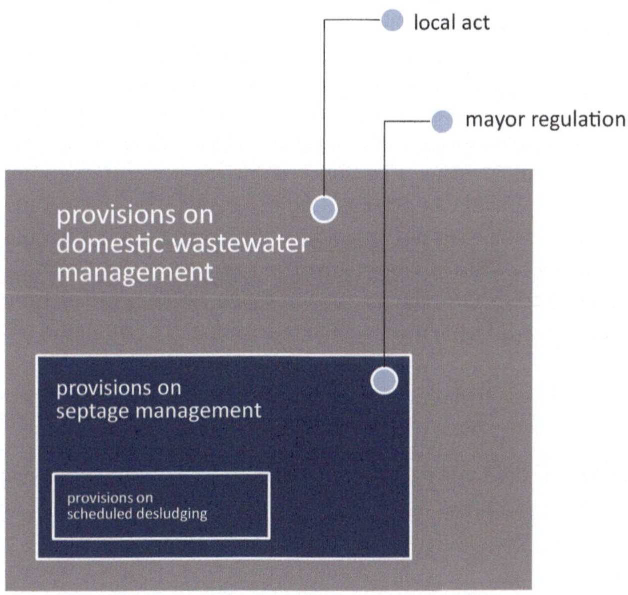


**Figure 8.3** Regulation of septage management should be part of the more comprehensive domestic wastewater management provisions. Hierarchically, the provisions for wastewater management should have a higher regulatory status while the provisions for the septage management are placed in the city government's regulation.

diagram, regulations at this level need to include provisions relating to institutional roles, private partnerships and payment mechanism for tariffs.

The principles and general provisions should be included in a regulation at the level of local act or local law, that is, the scope of wastewater management, types of services, role of local government, responsibilities of each party, etc. In the local act, the municipality is ordered to provide domestic wastewater management services while households are instructed to manage the wastewater they generate. Local act should contain various legal consequences of violating the provisions, that is, warning, fines and confinement.

As mentioned earlier, we should make provisions regarding tariff rates or septage service accounts in a separate regulation with a lower hierarchy. This needs to be done to anticipate adjustments of tariff in the future. It is very practical if the provisions on tariff are contained in the regulation at the level of the decision letter of the director of the scheduled desludging service provider. This is certainly possible if the city government or regent regulation states that the authority to determine the tariff is transferred to the director of the service provider.

## 8.4 FOSTER COMPLIANCE

Domestic wastewater management regulations in several countries include various types of legal consequences that can be imposed to those who violate the provisions. That includes written warning, rehabilitation order, fines, penalties and imprisonment. However, the city still needs a strategy to prevent households and building owners from violating provisions of domestic wastewater management; this is particularly to ensure households and building owners use proper septic tanks and maintain their performance by conducting regular desludging, or to connect to sewerage system.

Compliance strategy of a city might include:

- First, the municipality must inform the public about the obligation to manage their wastewater. Houses and buildings can use septic tanks as long as they conform with specification standard and periodic desludging requirement.
- Second, the municipality must ensure that the construction permit or building renovation permit are only granted after the proponent can present a proper plan of the septic tank construction.
- Third, the municipality needs to establish a septic tank registration and control system where each septic tank will have a certificate of conformance.
- Fourth, the municipality needs to conduct a regular septic tank inspection program to ensure that each septic tank is still in a good condition and has fulfilled its obligation of scheduled desludging (Figure 8.4).
- Fifth, the municipality needs to link the ownership of certificate of conformance of the septic tank with the payment of property taxes or with the building transaction process.

In addition, municipality should also prepare a financial or technical assistance program to help households to build a proper septic tank.

## 8.5 PROMOTE REGULATIONS

Municipalities often assume the public is aware of and has an understanding of their obligations in managing their wastewater. The fact is they have never heard of such a regulation. They do not know about the legal consequences of violating the provisions. To promote the obligations, the municipality needs to

**Figure 8.4** The municipality needs to regularly inspect every septic tank in its area, particularly to find out the condition, the utilization and the last desludging conducted. It is also necessary for the municipality to check on under-construction septic tanks.

develop a strong communication strategy which outline all activities and communication means to reach the main target groups.

A press conference at least should be conducted after a regulation is formalized. In addition, a series of seminars and public gatherings are made to ensure households are aware of their obligation to conduct periodical septic tank desludging. Flyers are made to be distributed by the municipality and community groups to households. More about the promotion is discussed in **Step 13: Promote service**.

# Step 9

# Involve partners

*The main service providers on scheduled desludging do not need to have their own fleet. There will always be parties ready to partner with the main service provider. They can take part in the work provided there is a mutually beneficial agreement. The financial and technical capacity of potential partners must be considered in the selection process. Ensure that the rights and obligations of each party will be clearly stated in a written cooperation agreement.*

## 9.1 OUTSOURCING BENEFITS ALL

There are several advantages if the main service provider outsources the operations of scheduled desludging fleets to operating partners. First, the main service provider will be released from the financial burden to buy desludging units and run a garage and maintenance facility. Second, the main service provider will also be released from the responsibility for the septage. The operating partner will handle the septage with care because any environmental pollution during transportation will cost them money and reputation. All will be provided by the operating partners, either from private sectors or community groups.

Other advantages of outsourcing scheduled desludging operation are:

- Increased responsiveness to operational fluctuations: An increase in the number of buildings to be served will increase the frequency of the desludging operations and the number of fleets. With an outsourcing scheme, a main service provider may get as many desludging fleets as they need from a partner. The main service provider does not need to worry about investment costs and the recruitment of additional crew members.
- Reduced labour liabilities: The private sector partner will be the party that recruits and manages the desludging fleet crew. All labour liabilities are with the partner, including financial liabilities. The main service provider can minimize the number of staff, limited to those who carry out managerial and supervisory functions.


doi: 10.2166/9781789062281_0061

**Figure 9.1** A private desludging firm will get benefits from their involvement as scheduled desludging operating partners. In addition to increased desludging work, the image of the company will also increase. The scheduled desludging scheme in Surakarta has involved a company called Daffa Jaya which now becomes one of the best desludging companies in the city.

- Better cash flow: Payment of service fees to private partners can be done regularly, either weekly, monthly, or bi-monthly. Daily operating costs will be borne in advance by the partner. This will help the cash flow of the main service provider.

The disadvantage of such outsourcing is the lack of control to how the scheduled desludging operations are being performed. Outsourcing partners are often motivated by profit rather than job quality. Work can be completed quickly, but it does not always meet the quality expected. Another disadvantage is the extra amount of money that must be paid to the partner as their fee. Although outsourcing work is generally considered cheaper, we must be aware of any potential of unexpected costs. For instance, due to extra work of fixing the septic tank hole and extra length of desludging hose required to reach a septic tank. There is also a concern about the improper use of personal data of houseowners by outsourcing partners. The main service provider should exercise extra caution before passing the data to a partner.

On the other hand, an outsourcing scheme offers several benefits for the operating partner. The main benefit is the guarantee of more frequent and regular desludging work (Figure 9.1). Accordingly, they will get a more sustainable income. Because of their direct engagement with the households, the partner will have greater opportunity to get additional work. For example, septic tank repairs and on-demand desludging later.

## 9.2 SELECT A RIGHT PARTNER

The main service provider must conduct an open selection process to appoint operating partners. Prior to that, the terms of reference of the work must be established that will be outsourced. There are at least five steps that need to be taken to select an operating partner (see Figure 9.2).

The selection process should include all the firms and entrepreneurs in the city. Selection criteria must be informed to all potential firms and entrepreneurs before the selection process which may include (a) being a legal entity and having a business license, (b) having a desludging unit in accordance with specification

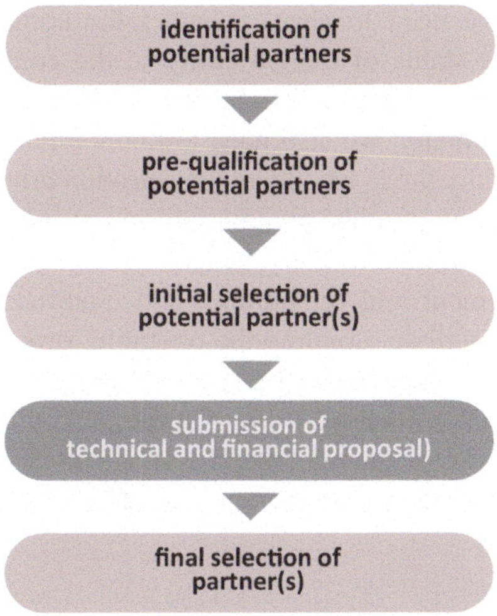


**Figure 9.2** The selection of scheduled desludging operating partners begins with the identification of the potential partners. After selecting several candidates who meet the criteria, potential partners are given an equal opportunity to submit technical and cost proposals. All proposals must be reviewed in the same way for comparable results. Scheduled desludging operating partners are then determined openly and fairly.

standard and (c) having trained crew members. Identifying potential operating partners in a city can be done in various ways. Start by looking at the operational records of a septage treatment facility. They must have a record of desludging firms and entrepreneurs who dispose the septage into the facility. Another way is to see desludging service advertisements which can be found on the side of the road, in newspapers and on the internet (Figure 9.3).

**Figure 9.3** Information on desludging services can be found in various media, especially on the internet. Some trucks also print the company name and telephone number on the tanks of their desludging trucks.

At the prequalification stage, we need to check whether the company has fulfilled the announced selection criteria. Starting from the status of a legal entity to the competence of its workers. It is also important to know how long they have been in this business and to know the number and condition of vacuum units they currently have. Finally, we also need to know their motivation to become partners for scheduled desludging operations. Only those who meet the selection criteria can continue with the process.

Those who pass the prequalification must be given a fair opportunity to submit technical and financial proposals. The technical proposal outlines (a) the type and number of desludging trucks to be used, (b) safety and other supporting equipment and (c) number and qualification of staff to assign. It is also important that a technical proposal will also outline the operating procedures which will later be applied as well as safety procedures. On the other hand, the financial proposal should state the service fees and payment terms proposed by potential partners. Ensure that the procedure for selecting operating partners is in accordance with formal procurement requirements. If all processes are carried out correctly, the main service provider will have a legitimate, competent, and affordable operating partner.

## 9.3 AGREE TO RIGHTS AND OBLIGATIONS

Both parties must sign an agreement on the outsourcing of the desludging fleet. We need to ensure that the agreement will include (a) the identity of the parties to the agreement, (b) intentions and objectives, (c) scope of cooperation, (d) technical requirement and specifications, (e) timeline, (f) rights and obligations, (g) wages or compensation for services and methods of payment. A cooperation agreement usually also includes other provisions, such as work safety, supervision and control of work, termination of the agreement, and dispute resolution. The following table lists some of the obligations to be included in the cooperation agreement between the main service provider and the operating partner. **Annex D** shows an example of outsourcing agreement between the main service provider with a private company for desludging work in the city of Surakarta (Indonesia).

Obligations of the main service provider and operating partners.

| Party | Responsibilities |
|---|---|
| Main service provider | (1) Inform building owners of plans for scheduled desludging.<br>(2) Provides the task of desludging with certain frequencies.<br>(3) Ensure that the septage treatment plant is ready to receive and treat septage.<br>(4) Monitor operation of the partner.<br>(5) Pay compensation according to the performance of operating partner. |
| Operating partner | (1) Uses desludging units and fleet crews according to requirements.<br>(2) Perform standard operating procedures.<br>(3) Conduct desludging according to the assignment.<br>(4) Transport septage to the specified septage treatment plant.<br>(5) Create and submit task implementation reports to the main service provider. |

The main service provider can also require the operating partner to perform other additional tasks, such as submitting desludging service bills, assessing the condition of septic tanks in the buildings and houses.

# Step 10

# Prepare the fleets

*Scheduled desludging operation depends mainly on the desludging fleets. Not only on the vehicles, but also on the crews and operational procedures. Fleet performance determines the level of public satisfaction with scheduled desludging. The appearance and cleanliness of the desludging vehicles will affect the public image on the scheduled desludging. The latest tracking technology needs to be used to monitor the vehicles.*

## 10.1 NO MORE HANDCART

Given the high frequency of desludging and the distance that must be travelled, scheduled desludging operation needs to use a motorized vehicle equipped with sludge tank and vacuum pump. No handcart and no manual pumping should be used in a scheduled desludging operation (Figure 10.1).

Several factors must be considered when we choose desludging unit, the factors include (1) width of the road to be passed, (2) travel distance, (3) traffic conditions, (4) required volume of sludge tank, (5) maximum gross weight limits of the truck, (6) comfortability and safety, (7) price, (8) ease of maintenance and (9) availability of spare parts. The implications of desludging vehicle cost on the financial performance of the scheduled desludging scheme need to be carefully assessed. Most of the scheduled desludging operation cost are for operation and maintenance of the desludging vehicle.

The choices of motorized vehicles that are worth considering for desludging operations are (1) three-wheeled motorized carts, (2) pick-up cars and (3) trucks. It is good if the scheduled desludging operation is equipped with all types of vehicles, given the diversity of road widths and traffic conditions as well as volume of septic tanks. For buildings in areas with narrow access roads, desludging operations inevitably need to use motorized carts or pick-up cars (Figure 10.2).

Sludge tanks can be manufactured for different volumes. From 0.6 $m^3$ for motorized carts to 12.0 $m^3$ for large trucks. Consider the desludging and transportation mode before we determine the tank size, especially if the scheduled desludging operation will apply a fixed volume desludging mode (see **Step 6: Design**


doi: 10.2166/9781789062281_0065

**Figure 10.1** A fleet of scheduled desludging consists of three components, namely (1) desludging units, (2) crew members, and (3) operating procedures. Collective compatibility of the three components will make scheduled desludging operations run efficiently, effectively, quickly, cleanly, and safely. The good appearance of these three components will increase public trust toward scheduled desludging service in the city.

**Figure 10.2** The most important parts of a desludging unit are (1) sludge tank, (2) vacuum pump, and (3) desludging hose. Each unit must also carry safety equipment and first-aid kit box.

**operations**). If we apply a fixed volume desludging of 1.5 $m^3$, a 3-$m^3$ tank desludging truck will be able to serve two households before leaving for the septage treatment plant. With a 4.5-$m^3$ tank, a desludging truck will be able to serve three households before leaving for the septage treatment plant. In addition, consider also the road width and weight limits before we determine the size of the tank. The bigger the sludge tank, the

**Figure 10.3** Sludge hose is generally made of flexible PVC material with a size of 3 inches or 4 inches. Most trucks carry 30-m hoses except for special orders for homes that are far from the highway.

heavier the burden of the road posed by the desludging unit. In full condition, a desludging truck with a 3 $m^3$ tank volume will weigh around 5 tons (Figure 10.3).

## 10.2 TWO ARE ENOUGH

Two crew members are needed for every desludging unit. Collectively, these two individuals will carry out duties to:

(1) Determine the route; in order to obtain the most efficient travel plans according to the desludging work order.
(2) Ensure the desludging unit readiness; especially regarding truck engine, sludge tank, vacuum pump, sludge hose, safety equipment, first aid kit, GPS monitoring equipment and work forms.
(3) Drive the desludging unit; from the pool, the house or building where septic tank is used to the septage treatment plant and back to the pool or move for the next cycle of operation.
(4) Communicate with customers; starting from confirming customer address and willingness to accept the service, explaining customer's rights and obligations as well as work procedures to be performed.
(5) Check septic tank readiness and condition; including making sure septic tank lid is open, measure the depth of settled sludge, record conditions and take photos of septic tank.
(6) Install sludge hose; to connect the septic tank with the truck's sludge tank.
(7) Control vacuum pump operation; ensuring it will work according to the needs and conditions.
(8) Monitor volume of the contents of sludge tank; in order to maintain the accuracy of the amount of septage that has been pumped out.
(9) Record desludging operation; either in writing or digital recording including identification and address of customers, duration of desludging and septage volume.
(10) Monitor septage disposal; to ensure that all septage is disposed in the septage treatment plant according to the standard operating procedures (SOPs) and recorded by the plant operator.
(11) Ensure safety; make sure that safety procedures are applied and safety gears are used.
(12) Check documentation and prepare daily report.

One crew needs to have the skills and proper license to drive a desludging unit. The other crew member should have skills in assessing and emptying septic tanks. Other tasks can be divided between the two crew members. Tasks performed in and around the desludging unit should be given to the driver. For example, the task of controlling pump operation, keeping record of desludging operation and ensuring compliance with work safety requirements.

## 10.3 WORK BY PROCEDURES

The SOP must be adhered to by all scheduled desludging crew members to maintain quality of service, safety of crew members and homeowners, work efficiency, good documentation, administrative order and to prevent environmental impacts. An SOP must clearly outline the work steps that must be carried out by all desludging crew members to meet operating objectives and performance indicators. The contents of SOP must describe the purpose of the assignment, the scope and responsibilities, definitions and references, work flowcharts, estimated duration and documentation.

A scheduled desludging operation should follow a set of standard procedures which at least includes procedures for (a) pre-departure check, (b) septic tank emptying operation and (c)

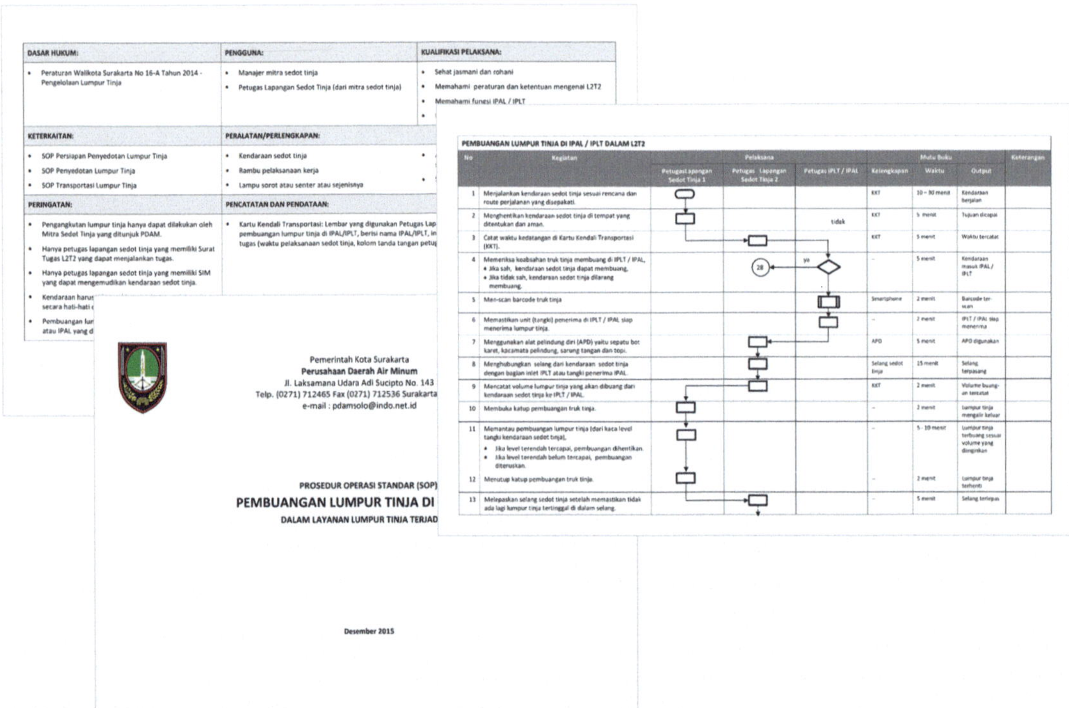


**Figure 10.4** There are many formats of SOPs and each country has its own standard format. However, each SOP format should be able to describe all required work steps sequentially. Performance indicator of each step must be stated along with the individual responsible for the step.

septage disposal operation. The details of each SOP may differ from one service provider to another but in general the contents will consist of the same set of generic activities (see the following tables) (Figure 10.4).

Generic steps of SOPs for pre-departure check.

| Steps | Description |
|---|---|
| 1. Daily briefing | The crew members discuss and determine (a) buildings to serve, (b) vehicles to use and (c) septage treatment plant to go. |
| 2. Customer confirmation | The crew member contacts the building owners to confirm (a) the address and route to the building, (b) there will be a person who would accompany the fleet and (c) the septic tank lid is open or will be open. |
| 3. Customer septage treatment plant | The crew member contacts septage treatment plant operator to ensure that the plant was willing and ready to accept the septage that will be brought by the desludging truck. |
| 4. Decide route | The crew member decides the most efficient and fastest route to reach the building. |
| 5. Inspect desludging unit | The crew member ensures the vehicle and pump are functioning properly, the fuel is available, sludge tank is not full, and the sludge hose is set. |
| 6. Check forms and equipment | The crew member checks the availability and functioning of communication equipment, safety gears, assignment sheets and forms, crew identification card. |

Generic steps of SOPs for septic tank emptying.

| Steps | Description |
|---|---|
| 1. Travel to building | The driver takes the desludging unit to the targeted building or house according to the specified time and route. |
| 2. Vehicle stop | The driver stops the desludging unit in a safe place (on a hard and level surface) which will not interfere with the traffic. |
| 3. Introduction | The crew member shows his/her identity card and explains the purpose of the visit, the rights and obligations of the building owner, procedures to be taken and expected cooperation from building owner. |
| 4. Use protective equipment | The crew member uses personal protective equipment (PPE) which consists of work clothes, rubber gloves, hats, protective glasses, rubber boots and mask. |
| 5. Observe the septic tank | The crew member checks whether the lid of septic tank is open and takes record of its condition, in particular regarding the depth of settled sludge and its hardness. |
| 6. Install hose | The crew member places the sludge hose in safe place between the desludging unit with the septic tank. If needed, the crew member should move and park the desludging unit in a closer spot. |

(*Continued*)

Generic steps of SOPs for septic tank emptying (*Continued*).

| Steps | Description |
|---|---|
| 7. Pump out septage | The crew member operates the pump and opens the tank valve so that the septage flows into the sludge tank until it reaches the desired tank height. |
| 8. Cleaning | The crew member cleans the area and ensures that all desludging activities do not leave any drops of dirty water and impurities in the property that may create odors. |
| 9. Tidying | The crew member cleans the hose and puts it back to its place in the desludging unit. |
| 10. Settlement | The crew member asks the building owner to sign the desludging duty sheet and give one copy to the building owner. |

Generic steps of SOPs for septage disposal.

| Steps | Description |
|---|---|
| 1. Trip to the treatment plant | The driver takes the desludging unit to septage treatment plant according to the specified time and predetermined route. |
| 2. Examination of sludge tank | The septage treatment plant operator checks the validity and eligibility of the desludging unit to dispose septage in the plant. |
| 3. Desludging vehicle stop | The crew member stops the desludging unit at the designated spot at the septage treatment plant area. |
| 4. Use protective equipment | The crew member uses PPE which consists of work uniform, rubber gloves, hat or helmet, protective glasses, safety shoes and mask. |
| 5. Install hose | The crew member connects the hose from desludging unit to designated receiving unit in the treatment plant. |
| 6. Drain septage | The driver opens the drain valve so that the septage can freely flow out to the receiving unit until the sludge tank is empty. |
| 7. Cleaning | The crew member cleans the area and ensures that their activities do not leave any drops of dirty water and impurities. |
| 8. Tidying | The crew member cleans the hose and puts it back to its place in the desludging unit. |
| 9. Settlement | The crew member asked the septage treatment plant operator to sign the septage disposal task sheet. |

## 10.4 KNOWING THE RISKS

Septic tanks are dangerous because they accumulate a variety of toxic gases. Methane can be very toxic in high concentrations while it is explosive in a concentration commonly found in septic tanks. Other gases are hydrogen sulfide, ammonia, nitrogen dioxide, sulfur dioxide, and carbon dioxide. Hydrogen sulfide in small

**Figure 10.5** Collaboration between the desludging fleet crew members is needed so that all tasks can be performed properly. Most importantly, the crew members must have good attitude and skills to communicate with customers.

concentrations may cause eye irritation, shortness of breath and unrelenting cough. Exposure to higher concentrations can be fatal quickly. Other potential hazards are suffocation, collapse of septic tanks and health risks from bacterial or viral infections. Deaths that involve falling into or accidentally entering a septic tank occur throughout the world. Septic tank explosions also occur frequently, especially as a result of careless work in desludging operations.

Crew members must be assured that their desludging work carries a high risk to health and safety. They also need to be aware that they or their co-workers can experience accidents or health problems due to their careless work. They must also be aware that residents of the house can become victims of their mistakes in work. It is strongly recommended to remind the health and safety risks for all crew members during the daily briefing.

The use of personal protective equipment (PPE) by desludging crew is a must. Each person working nearby the septic tank must use rubber gloves, protective glasses, safety shoes and masks (see Figure 10.5). Another prevention is to ensure that the crew will follow the SOP when opening the septic tank lid, checking septage depth, checking septic tank chamber, mixing septic tank content, pumping septage out, and closing the lid. Anybody should never be allowed to work alone at a septic tank because falling into a septic tank or inhaling methane gas can be fatal. They can also fall into the septic tank when they examine its interior. **Annex F** provides more information on the safety issues of scheduled desludging operation (Figure 10.6).

## 10.5 MONITOR THE OPERATION

Every action of the fleet needs to be monitored. It starts when the desludging unit leaves the garage, goes to the building, empties the septic tank, disposes septage at the treatment plant until it returns to the garage. The monitoring encourages fleet crews to work according to standard procedures and daily operational

**Figure 10.6** PPE that must be used by the crew of the desludging fleet, especially for those who are directly involved in septic tank emptying operation.

plans. The fleet will come to the building according to the schedule and dispose septage at the specified treatment plant.

Monitoring can be done manually, that is, by recording fleet activities at each stop, namely the garage, the building, the septage treatment plant and even the rest area. Information to be recorded includes (a) identification, name and address of the place, (b) time of arrival and departure, (c) activities carried

**Figure 10.7** The use of digital GPS tracking systems allows the scheduled desludging operation to be monitored in real-time and continuously. Each building has a barcode to be scanned by a smartphone carried by a crew member. Internet connection allows the scan results to be sent to the data center. The crew no longer needs to carry a control card and ask for a signature from the building owner. This system has been developed by IUWASH for use in scheduled desludging scheme of Surakarta.

out at each stop. The crew write down the activity on the operation control card. Before leaving each stop, the crew must ensure that each operation control card is signed either by the garage officer, building owner or treatment plant operator. The operation control card is periodically handed over to the supervisor of scheduled desludging operation.

With current advances in information technology, monitoring of the fleet can also be done digitally (see photo below). The global positioning system (GPS) tracking system has been used in many transportation companies. The tracking system can continuously find the location of a vehicle and then display it on a digital map that can be seen in real time by the supervisor. It provides visibility into the whereabouts of the fleets which makes the monitoring much more effective. Some GPS tracking systems can even recommend the optimal route to crew members which will allow them to reach their destination faster. All tracking data will be stored in a central computer system so as to enable the evaluation of the performance of each desludging fleet, including the desludging frequency, number of trips and working time (Figure 10.7).

The monitoring records can also be used to verify the work progress of a desludging fleet. If the main service provider outsources the desludging work to an operating partner, the data from the monitoring should be used to verify the claim in the invoice. Payment can only be made if the record data show that they have desludged septic tank according to the schedule and disposed the septage at specified treatment plant.

# Step 11
# Install management system

*A scheduled desludging scheme must use a management information system (MIS) that has the ability to collect, compile, record, send, and process data of customers, operation and finance in an integrated manner. MIS will ensure that all registered buildings can receive desludging service according to the specified schedule. The use of digital information technology enables the service provider to manage desludging operation for a high number of buildings and in real-time.*

## 11.1 ALL IS CONTROLLED

A scheduled desludging operation consists of dozens of activities carried out by different divisions within the organization of the service provider (Figure 11.1). Some activities may need to be carried out by others, such as by the operating partners and septage treatment plant managers (see Figure 11.2). Communication and coordination become critical in the scheduled desludging operation. Each party involved must work synchronized according to standard procedures and other agreements. The exchange of information between parties must run smoothly, and all information will be recorded to enable quick and accurate evaluation and verification of operating performance. Therefore, it is highly recommended that the scheduled desludging operation should be supported by a management information system (MIS) that utilizes information and communications technology.

An MIS is used for coordination, control, analysis, and visualization of information in a service provider. In full, an MIS in a scheduled desludging operation must be able to:

- store and sort data of buildings that use viable septic tanks,
- compile a list of customers who will get a desludging service in a time period,
- set weekly or monthly operating schedule of desludging,
- assign a desludging fleet according to the specified schedule,
- document desludging operation, septage transportation and septage disposal,
- create and print bills for the customers,


doi: 10.2166/9781789062281_0075

**Figure 11.1** Scheduled desludging service provider needs to use a MIS based on digital information technology. An MIS will help service provider to control the desludging, transportation and disposal of septage to meet the planned target and time.

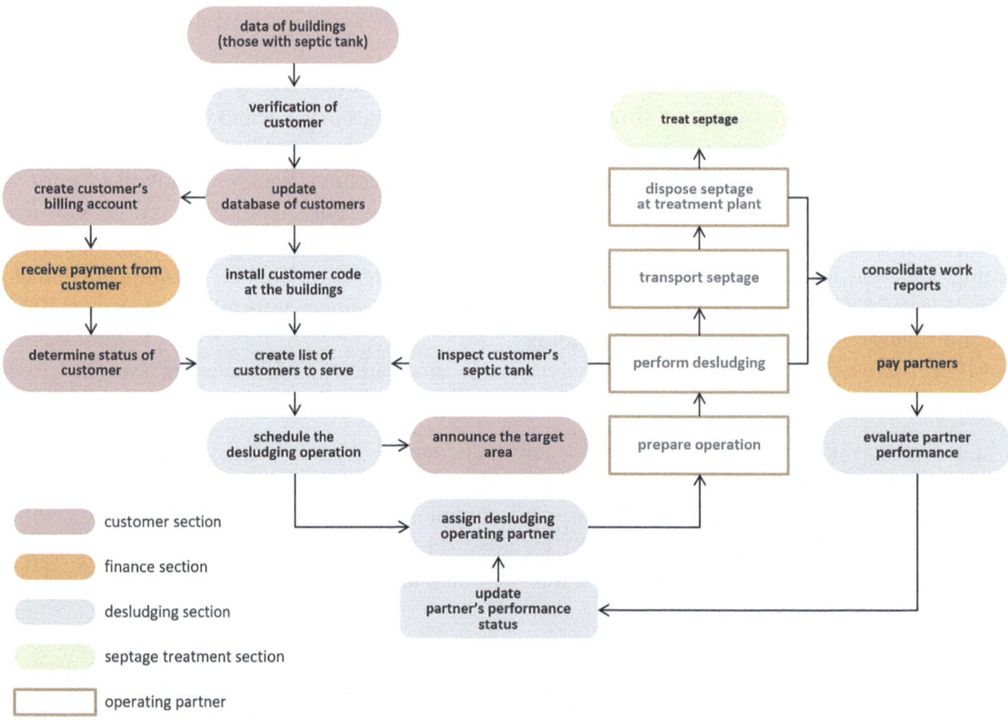


**Figure 11.2** Series of activities and information exchange traffic in a scheduled desludging operation. These activities involve several parties, both inside and outside the service provider organization.

- record payment transactions of the bills,
- calculate the amount of compensation to be paid to the desludging operating partners,
- prepare technical and financial performance reports.

An MIS can also be designed to assist a service provider in assessing their own performance and their operating partner's performance. For example, assessing the average time required for desludging, the number of desludging operations over a period of time. A good MIS should have the ability to sort out information of buildings who are viable and eligible for desludging service according to a set of agreed criteria. One of these criteria is the obedience of building owners in paying desludging service bills. An MIS will help the service provider to compile a septic tank desludging plan that mentions the identity of customers along with the address and desludging schedule. An MIS will also use customer information in the database to create service bills for each customer.

**Annex G** briefly describes an MIS that has been developed by IUWASH PLUS program to support scheduled desludging in several cities.

## 11.2 RELY ON DATABASE

An MIS of a scheduled desludging operation relies on the database of buildings that use septic tanks. The customer database must at least contain information of (a) identity of buildings or customer, that is, number, name, address, telephone number and customer classification and (b) condition of the septic tank, namely location, accessibility, volume, source of waste and the last desludging year. If it is made consistent with the questionnaire used for surveying the septic tanks (see **Step 5: Assess targets**), we can use data collected to fill the database (Figure 11.3).

The database can be specially designed to have a data structure suitable for a scheduled desludging operation in a city. Creating a new database of buildings or customers may not be necessary if the desludging operation will involve an existing institution that already has a database for their customers. For example, a water utility that already has a database of their water customers. However, before it is used, we need to carefully study the existing database structure to make sure that the record type and

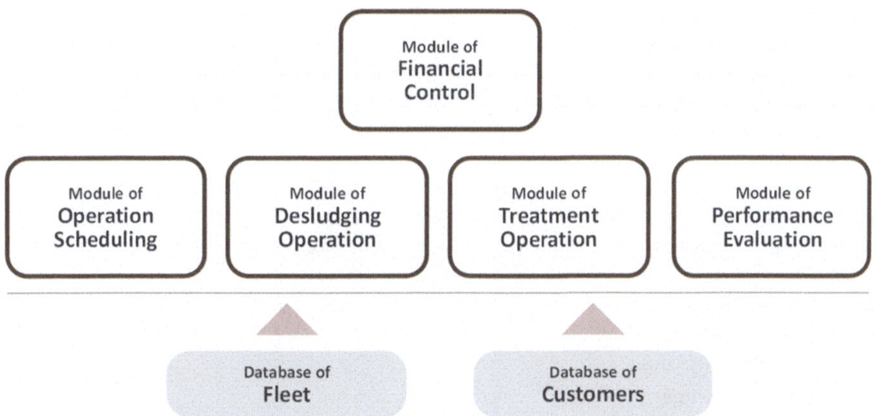


**Figure 11.3** The MIS for scheduled desludging operation may consist of a set of modules to specifically support administrative, technical and financial processes of their operations. The core of an MIS is the customer database that will underlie other functions.

field type are compatible with the specific needs of a scheduled desludging operation. It is almost certain that the database of water utility should be improved to include data field for septic tank.

An MIS should also have database of the fleets, both for the desludging units and the crew members. A database of desludging unit at least needs to contain information of (a) identification and license numbers, (b) type, brand and year, (c) tank volume, (d) gross weight, (e) operating records, and (f) maintenance records. A database of crew members might include information of (a) name and birthdate, (b) identification number, (c) job title, (d) address and contact number, (e) license number, and (f) operating records.

## 11.3 CONNECTING THE PARTNERS

The MIS is designed to support information exchange between divisions in service provider organization, particularly between customer, technical and financial departments. The exchange of information between these three divisions is necessary for the preparation of the desludging plan. If a service provider does not want to serve buildings that have not paid the service bill, the finance department must inform the technical department to prevent desludging service to those buildings. The MIS then does the work and fleet crews are informed through their smartphones.

### MIS for scheduled desludging in surakarta (indonesia)

The scheduled desludging scheme in the city of Surakarta (Indonesia) is run by the local water utility, that is, Perusahaan Daerah Air Minum (PDAM) Surakarta. They utilize an IT-based Management Information System (MIS) which consists of three parts, namely (see the diagram):

- The main MIS unit (with website) that is used by the controller at the PDAM.
- The treatment mobile MIS unit to be used by the treatment plant officer to record and report the frequency and volume of septage disposal.
- The desludging mobile MIS unit to be used by the fleet crew to record and report the desludging and transportation operations.

Based on customer database, PDAM makes a list of buildings to serve. The list is submitted to the fleet crew via the desludging mobile system. When performing desludging operation, the fleet crew will scan the customer's barcode and send information to the main MIS unit. Septage treatment plant officer will use their mobile MIS unit to record and send information on septage disposal to the main MIS unit. Treatment plant officer must scan the barcode in desludging truck.

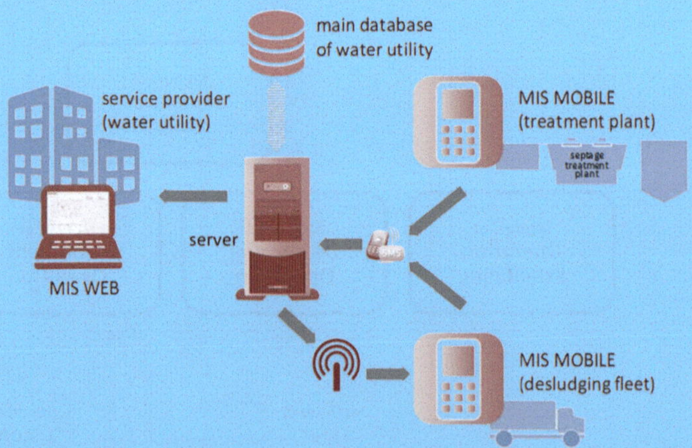


All information is collected on the server, while the PDAM controller can monitor it through the MIS Web.

The main MIS unit operated by the PDAM controller has modules of:

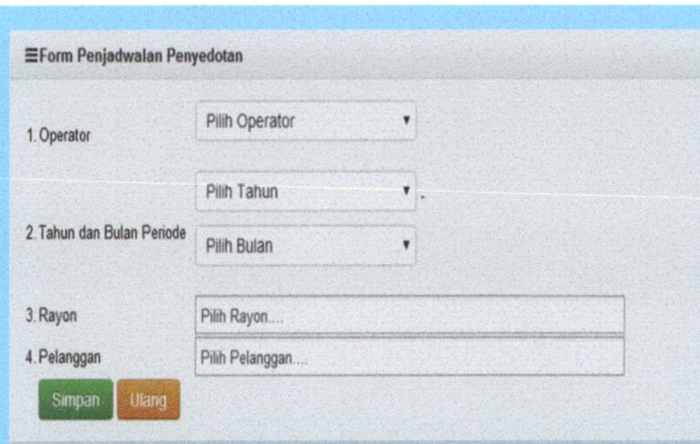


- Fleet data module; to manage the data of desludging fleets.
- Request module; to manage requests of buildings on desludging outside the scheduled scheme.
- Scheduling module: to set the schedule of septic tank desludging which is based on service zones, building types or both (see picture in the right).
- Dashboard; to see the performance of desludging operations, such as locations, desludging monthly data and graphics on average septage disposal.
- Operation monitoring module; to monitor the position and time of desludging fleet activities.
- Financial reporting module: to see the status of the invoice and the amount that must be paid by a customer.
- Service billing module; to integrate billing data of desludging service with the water supply billing system.

The PDAM Surakarta uses an Intel Core i5-4590 desktop computer to accommodate the main MIS unit. Meanwhile, the fleet crews and septage treatment plant officers use Android-based smartphones to operate the mobile MIS units. Overall, the MIS of PDAM Surakarta has eight main application modules and three mobile application modules.

# Step 12

# Plan finance

*Now it is time to determine the tariff for the scheduled desludging service for households and other targeted buildings. Financial projections are made to ensure the proposed service tariffs will not create an unhealthy financial condition of scheduled desludging scheme. The municipality can set the tariff of scheduled desludging to be billed in monthly instalments by building owners, but there are other payment schemes to be considered.*

## 12.1 EACH HAS ITS OWN TARIFF

The basic tariff for scheduled desludging is usually presented as the rate imposed on households. Tariffs for other building classification are usually different. Commercial buildings generally have higher tariffs. They have larger septic tanks compared to those in households. Obviously, the desludging fleet must spend more energy and time to pump out and transport septage from those buildings. The use of more resources must be compensated by the imposition of a higher tariff to commercial buildings and public offices. If we have the base tariff for a household, we can determine the rate of other building classification by using the ratio of septic tank volume among different building classifications. The following formula can be used to determine the tariff:

$$\mathrm{BTC} = \mathrm{BTH} \times \frac{(\mathrm{STVC})}{(\mathrm{STVH})}$$

in which BTC = basic tariff for commercial buildings, BTH = basic tariff for households, STVC = septic tank volume in commercial buildings, and STVH = septic tank volume in households.

Information about the average volume of septic tanks for each customer classification can be obtained from the septic tank survey (see **Step 5: Assess the targets**). The following table illustrates the results of the calculation of the basic tariff for each customer classification, assuming we apply the fixed volume desludging mode.


doi: 10.2166/9781789062281_0081

Illustration of tariff calculation for each building classification.

| Classification | Average Septic Tank Volume ($m^3$) | Desludging Volume ($m^3$) | Tariff (USD/month) |
|---|---|---|---|
| Households | 2 | 1.5 | 0.9 |
| Commercial | 8 | 6 | 3.6 |
| Public office | 6 | 4 | 2.4 |
| Social | 4 | 3 | 1.8 |

There are several things that may influence the determination of actual tariff of the scheduled desludging service, that is

- building size, which makes large buildings pay higher tariffs than small buildings,
- profit target, which applies in a service provider with status as regional company while those with status as municipal agency may not need to consider any profit in their tariff calculation,
- cross-subsidy policy, which makes commercial buildings pay higher to cover some of the operational expenditures in serving poor households or social buildings (Figure 12.1),
- tariff for other wastewater services, the scheduled desludging tariff should not be much different from the on-demand desludging tariffs to avoid opposition from public or private desludging companies.

**Figure 12.1** Commercial buildings are considered to have a higher economic value than households. It is common for a city to charge commercial buildings with a higher tariff than households, public offices and social buildings. This principle also applies among households of which luxurious houses are subject to higher tariff than simple houses. Or houses in elite areas that must pay higher tariff than those in common areas.

Finally, the political considerations of city government and legislators will affect the actual tariffs. Many city leaders do not want to impose the actual tariffs or even the cost-recovery tariffs on wastewater services on their people for political reasons. The following table shows the tariffs of scheduled desludging service in the city of Surakarta (Indonesia).

Scheduled desludging tariff in Surakarta (Indonesia).

| Classification | | Tariff |
|---|---|---|
| Households | Very low income | IDR 5000 |
| | Low income | IDR 8000 |
| | Medium income | IDR 8500 |
| | Luxury | IDR 9000 |
| Commercial | Small | IDR 45 000 |
| | Large | IDR 70 000 |
| Public buildings | Government offices | IDR 17 500 |
| Social | Schools | IDR 17 500 |

**Note:** 1 USD = IDR 14 000 (in December 2019).

Some cities consider eliminating tariffs for very low-income households. If this is done, the municipality must compensate it by raising the tariff for other groups. It is important at the end that all the tariffs on average still have cost recovery.

## 12.2 MANY PAYMENT OPTIONS

In the previous descriptions, scheduled desludging tariffs are calculated with the assumption that the payment will be made in monthly instalments (see **Step 3: Make the initial concept**). It might be convenient for building owners but paying in monthly installments can create difficulties for the service provider. Creating and sending monthly bills are troublesome. In addition, there is a high possibility that under a monthly installment option a service provider will have to pay a portion of the desludging costs in advance. A monthly installment option is more applicable in a scheduled desludging scheme that is managed by a water utility. They can combine the scheduled desludging bill with the monthly water bill that already has a regular delivery system. The following table shows several other options that should be considered as payment methods for scheduled desludging customers.

Options of scheduled desludging tariff payment.

| Factor | Payment Options | | |
|---|---|---|---|
| Time | 1. | Prepaid | Customers begin paying without waiting for services to be obtained |
| | 2. | Post paid | Customers begin paying after obtaining service |

(*Continued*)

Options of scheduled desludging tariff payment (*Continued*).

| Factor | Payment Options | | |
|---|---|---|---|
| Method | 1. | Combined | Customers pay scheduled desludging bill that are incorporated in the bill of other service, such as water supply, electricity, or garbage services. |
| | 2. | Single | Customers pay scheduled desludging bill, without being combined with the bill of other service. |
| Frequency | 1. | Instalments/ gradual | Customers pay scheduled desludging bill in instalments, either monthly, quarterly, six-monthly or annually |
| | 2. | All at once | Customers pay scheduled desludging bill at once. |

We can apply several payment options in one scheduled desludging system. Payment option for one customer classification may differ from other classifications. A one-time payment option can be applied to public offices, considering that the payment will be made by using municipality budget. Installment payment option that is billed in a joint bill with water supply service is appropriate to be applied for households. Monthly installments are more appropriate for low-income households. Meanwhile for middle–upper households, particularly in a single billing method, the payment at once will be more suitable.

Scheduled desludging payment can be made in cash (cash) or non-cash. In addition to regular payment points of the service provider, cash payments should involve payment agents in the service zones. Non-cash payment, or e-payment, is a way of making transactions or paying for services through an electronic medium. Popularity of e-payment systems have improved over the past few decades due to the increasing spread of internet-based banking and shopping. Non-cash payment avoids direct cash transactions between fleet crew members and building owners (Figure 12.2).

## 12.3 AVOID FINANCIAL LOSS

We need to carry out financial projections to ensure the proposed tariff rates will be sufficient to cover all scheduled desludging operating costs. To the extent possible, these tariffs can even provide substantial profits for the city and the service provider. We make financial projections based on some values used in

**Figure 12.2** Payments through ATMs have been made for various purposes, including payment for electricity services and drinking water. Its application for scheduled desludging is very possible, especially in big cities.

the scheduled desludging operation design (see **Step 6: Design operations**). If the financial projections show financial loss, then we can change some operating parameters and simulate the operation design again. However, it should be noted that profits may not come in the initial implementation years.

Scheduled desludging financial analysis is conducted with a set of assumptions as follows:

- Investment in infrastructure is borne by the government and other parties, so that the service provider is not charged with capital costs.
- Septic tank desludging operations are fully outsourced to private fleets.
- Tariff are fixed for an operating cycle (@ 3 years).
- Billing efficiency will not reach 100%.

The revenue comes only from desludging bill payments, after considering the level of billing efficiency. Financial projections are conducted for several scheduled desludging operating cycles. The following table shows the conclusion of the financial projection.

Financial projection of the scheduled desludging operation in Surakarta (Indonesia).

| Item | Unit | Cycle 1 | Cycle 2 | Cycle 3 |
|---|---|---|---|---|
| Operating revenue | IDR million/year | 7.825 | 11.508 | 16.751 |
| Operating expenses | IDR million/year | 6.772 | 8.598 | 10.960 |
| Profit before tax | IDR million/year | 1.053 | 2.911 | 5.791 |
| Note: | | | | |
| • Number of customers | Households | 61.745 | 69.149 | 79.534 |
| • Septage volume to desludge | $m^3$ / day | 129 | 151 | 174 |
| • Frequency of desludging | Per day | 69 | 79 | 89 |
| • Desludging truck | Unit | 12 | 12 | 13 |
| • Treatment facilities | Unit | 3 | 3 | 3 |

**Note:** 1 USD = IDR 14 000 (in December 2019).

**Annex H** contains examples of financial calculation spreadsheets.

# Step 13
# Promote service

*Scheduled desludging by law must be carried out to all septic tanks. With or without promotion, scheduled desludging will be demanded by buildings. Nevertheless, the promotion of scheduled desludging is still important. By knowing its reasons and benefits, the people will better support scheduled desludging. It is recommended that the scheduled desludging service should have a unique brand and building owners must know their rights and obligations related to scheduled desludging. Various ways of promotion need to be applied to reach diverse customer groups.*

## 13.1 PREPARE PROMOTION

In this step, we already know the specification of scheduled desludging service to be provided by the municipality, who and where the service targets are, and how much the tariffs will be imposed to house and building owners. In accordance with the four Ps marketing mix principle, the final principle to implement is the promotion. Although it is a mandatory action, we have to promote the scheduled desludging service to the owners of buildings that use septic tanks. Six steps may be needed to prepare the promotion of the service (see Figure 13.1). These generic steps can be modified according to the conditions and readiness of the municipality or the service provider.

Scheduled desludging is different from the existing on-demand desludging service. Its characteristics are mandatory, periodic, scheduled, professional, and inexpensive. This is the main characteristic of scheduled desludging that needs to be informed to the target group.

## 13.2 UNDERSTAND THE TARGET

The target group(s) of the promotion program is obvious, that is, the owners or residents of buildings that use septic tanks. If we want to be more specific on the target, we can choose households as the main target. Not just all types of buildings, but only households. A market research should be conducted to find out the level of knowledge, attitude, and habits or practices of households, especially in managing their wastewater.


doi: 10.2166/9781789062281_0087

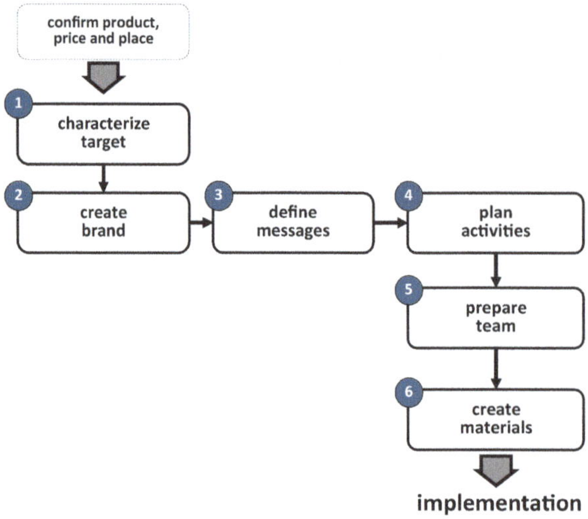


**Figure 13.1** Six generic steps to prepare promotion of scheduled desludging service in a city.

Research is also conducted to estimate the response of households to the upcoming scheduled desludging scheme (see the following table).

Character of the households will determine the messages to be conveyed in the promotion program and how they are conveyed. Some information about households needs to be obtained and studied before then concluded as the collective level of knowledge, attitude, and practices related to (a) use of septic tank, (b) desludging of septic tank, and (c) information exchange. Household surveys may also collect data on the average age, level of education, religion, and economic conditions of the homeowners. The results of the research can provide an initial indication of household interests in scheduled desludging service. If it turns out that the majority of households are not yet interested in scheduled desludging service, the promotion program needs to prepare and deliver information about the existence and benefit of the service.

Information for developing a promotion plan.

| | **Information** | **Use** |
|---|---|---|
| 1 | The level of household knowledge on regular desludging | Determine goals and promotional messages |
| 2 | Interest of household utilizing scheduled desludging service | Determine goals and promotional messages |
| 3 | Age and religion of homeowner | Determine promotion approaches and messages |
| 4 | Home ownership status | Determine promotion approaches and messages |
| 5 | Social activities in residential areas | Determine how to promote |
| 6 | Information sources | Determine how to promote |
| 7 | Access to social media | Determine how to promote |

If the information is still inadequate, special interviews can be conducted with community leaders in the target area. Village or community leaders or administrators, teachers, youth group leaders, and religious leaders can usually provide good information. They recognize the character and habits of the surrounding households, as well as knowing what efforts have been made to improve awareness on sanitation in the area.

In addition to households, promotion should also be done for secondary target groups which are expected to influence the perception of the households. Some examples of secondary target groups are community leaders, religious leaders, village officials, neighborhood officials, mass media, and representatives of government agencies.

## 13.3 GIVE A UNIQUE BRAND

Scheduled desludging service must have a unique identity. The use of relevant and attractive names, logos, and taglines will make scheduled desludging more recognizable to the public. All scheduled desludging logos and other identities must remain relevant to the type of service provided and represent the identity of the city. The goals of the branding will be achieved if the brand name, logo, tagline, and color continue to be used in all objects and activities used by the scheduled desludging operation (Figure 13.2).

A good brand name should be easy to pronounce, identified and memorized, and may as well give an idea about product's qualities and benefits. Avoid using a brand name that use the words of 'feces', 'desludging,' and 'wastewater' directly. Additionally, a logo also should be introduced to help customers recognize the scheduled desludging service. Logos are easier to remember than brand names because they display identity visually. For example, it will be easier if we use illustrations of healthy and good environmental conditions.

To create awareness about scheduled desludging scheme, it is important to apply the brand consistently. This includes its application in various promotional channels such as advertisements, word of mouth

**Figure 13.2** The desludging unit and its crew members are the face of the scheduled desludging scheme. If it looks modern and clean, people will have more trust in the service. The public will immediately understand that scheduled desludging provides a different service than the on-demand service.

publicity, social media involvement, sponsorships, and events. Strong brand awareness can be considered as a means through which customers become acquainted and accustomed to scheduled suctioning. Brand guidelines that function to maintain brand consistency are useful for establishing rules for creating a unified identity when connecting several elements in a brand such as colors, logos, and typography.

To create awareness on scheduled desludging scheme, it is important to apply the brand consistently. It includes use of various renowned channels of promotion such as advertising, word of mouth publicity, social media engagement, sponsorship, launching events, etc. Strong brand awareness can be regarded as means through which costumers become acquainted and familiar with scheduled desludging. Brand application guidelines will be very useful to maintain consistency in the application of elements of service identity in various media, such as colors, logos, and typography.

## 13.4 MESSAGES ACCORDING TO TARGETS

Scheduled desludging promotion must be designed to effectively and efficiently deliver awareness and knowledge-raising messages related to septage management to households. Packaging information must be tailored to the characteristics of the households, that is, level of education, perception, economic status, and preferred communication channels. The households should know its mandatory nature, form and scope of service, desludging period, differences with on-demand service, service benefits and processes. The features of the scheduled desludging fleet also need to be informed to the public, especially their modern, clean and professional look. The messages that need to be conveyed include:

- Reasons for the need of periodic desludging.
- Periodic desludging is required by law.
- All septic tank users must accept and cooperate with scheduled desludging service.
- Rate of service fee and payment procedure.
- Identity of service provider.
- Homeowners rights and obligations.
- Punishment for violating the scheduled desludging provision.

In addition to the messages above, we also need to include a variety of important information to change people's perceptions that are less precise about septic tanks and septage. For example, perceptions that there are no problems with the groundwater quality in their neighbourhood, a true septic tank is a tank that is never full and does not need to be drained.

## 13.5 USE ANY POSSIBLE MEANS

As with the promotion of other products, scheduled desludging promotion requires a set of above-the-line and below-the-line activities. Above-the-line activities such as advertising on radio, television and local newspapers are needed to introduce and echo the scheduled desludging service to wider communities. While the below-the-line activity is needed to explain the procedure and encourage the target groups directly to accept scheduled desludging operation (Figure 13.3).

Promotion activities, both above the line and below the line, which are worthy of being used to promote scheduled desludging include:

- Installation of banners in strategic locations that are visited or passed by people see (Figure 13.2).
- Installation of posters at public offices and exhibitions.
- Distribution of leaflets at promotional events or sent by post (see Figure 13.4).
- Sticker attachment at motorized vehicles and buildings.

**Figure 13.3** Promotional activities are generally costly. It would be better to focus on promotional activities only in designated service areas. For example, through the installation of banners at the village office where scheduled desludging will begin.

**Figure 13.4** Examples of leaflets used by PDAM Surakarta to introduce scheduled desludging to their customers. Use attractive visual appearance and simple language in the leaflet.

- Public service announcements on local newspapers, radio, and television.
- Talk shows on radio or television presenting government and service provider officers to discuss the scheduled desludging with the audiences or listeners. Village gathering where government and service provider officers introduce and discuss the scheduled desludging service with representatives of homeowners.
- Citizen meetings; both at special events and at community gathering events.

Promotion requires more than one type of media. Another media that we should consider using is the digital media, both social media and instant messaging applications. Digital promotion offers convenience, speed and flexibility in delivering messages. If there are disadvantages, digital promotion can only reach households that have internet access and use the same media applications.

The packaging of promotional message must be made according to the characteristics of each application. Packaging messages through photos and videos is more conveniently delivered through Instagram which is basically a visual platform. Public announcements with long narratives are inappropriate for Instagram. We better use emails for that. A short narrative (<160 characters) is better to be delivered via Twitter. The combination of these packages can be delivered via Facebook. Instant messenger applications such as WhatsApp and Line can contain long narratives accompanied by short photos and videos. Service providers can use the hashtag feature on Twitter and Instagram to attract the attention of more users of the application.

## 13.6 INVOLVE OTHER PARTIES

The municipalities need to be involved in scheduled desludging promotion. After all, the scheduled desludging service is a response to the obligation to periodic desludging by the municipalities. The participation of government officials will also add credibility to scheduled desludging in the eyes of the households. Cooperation with other parties should also be considered. We can collaborate with local mass media to intensify above the line promotion, including cooperation in advertising, advertorial, coverage, or interactive programs on scheduled desludging topic.

As for below the line activity, we can involve a local partner. They are the ones who know the condition of the people in the area. They will also be more easily accepted by the community. Examples are community groups, women groups, builders/small contractors, and desludgers.

# Annexes

(A) Assessment Form of City Strengths for Scheduled Desludging
(B) Questionnaire for Septic Tanks Assessment
(C) Draft of Mayor Regulation on Septage Management
(D) Private Sector Partnership Agreement for Scheduled Desludging
(E) SOP for Septic Tank Desludging
(F) Safety Issues of Scheduled Desludging
(G) MIS for Desludging Service by IUWASH PLUS Program
(H) Spreadsheet for Septage Desludging Financial Calculation


doi: 10.2166/9781789062281_0093

# ANNEX A

# ASSESSMENT FORM OF CITY STRENGTHS FOR SCHEDULED DESLUDGING

## ASSESSMENT FORM

### BASIC INFORMATION

1) Name of city: ____________________

2. Area: ____________ $km^2$/ha

3. Population: ____________ Persons

4. Population density: ____________ Persons per $km^2$/ha

### SANITATION CONDITION

1. Access to toilet: ____________ % population/household

2. Use of septic tank: ____________ % population/household/area

3. Sewerage system:
   - ☐ Yes, coverage ______ % population/household/area
   - ☐ Planned, for coverage ______ % population/household/area
   - ☐ None

4. Desludging service: ☐ Yes, by the municipality / others ☐ None

### DESLUDGING

2. Mandatory requirement:
   - ☐ Yes, ____________________
   - ☐ None

3. Operation mode:
   - ☐ On-call or on-demand basis
   - ☐ Scheduled desludging

4. Service provider:
   - ☐ Municipal agency, ____________________
   - ☐ Private firm, ____________________

☐ Individual, ______________________________

☐ Others, ______________________________

5. Payment of service: ☐ Before service

☐ After service

☐ Others, ______________________________

6. Payment point: ☐ On the location

☐ Office of service provider

☐ Others, ______________________________

7. Service charge: USD ______________ Per $m^3$/ operation / month / year

8. Influence by distance: ☐ Yes

☐ No

7. Facilities:

- Desludging truck: ________ unit, capacity____ $m^3$

________ unit, capacity ____ $m^3$

- Desludging tricycle: ________ unit, capacity ____ $m^3$
- Desludging cart: ________ unit, capacity ____ $m^3$

8. Performance

- Number of trips: ________ trips per day
- Households served: ________ households per day
- Septage volume: ________ $m^3$/day

9. Disposal of septage: ☐ Treatment facility ☐ Lake

☐ River or gutter ☐ Open field

10. Treatment fee USD ______________ Per $m^3$/disposal

11. Operating license ☐ Yes, ______________________________

☐ None

12. Distance

- Farthest: ____________ km
- Closest: ____________ km

- Average ____________ km

**SEPTAGE TREATMENT FACILITY**

1. Location: ________________________
2. Built in year: __________
3. Total area of land: __________ $m^2$/ha
5. Treatment unit: ☐ ________ ☐ ________ ☐ ________ ☐ ________ ☐ ________ ☐ ________
6. Receiving water ☐ River ☐ Lake
7. Condition ☐ Good ☐ Damaged ☐ Not operational
8. Design capacity ______ $m^3$/day
9. Used capacity ______ $m^3$/day
10. Number of disposals: ______ Times/day
11. Reuse of sludge ☐ Yes, ________________ ☐ None
12. Operator: ______ Person

**INSTITUTIONAL FRAMEWORK**

1. Planning: ________________________
2. Construction: ________________________
3. Regulating: ________________________
4. Desludging: ________________________
5. Treatment: ________________________
6. Supervision: ________________________

**REGULATIONS**

1. Use of septic tank: ☐ Yes, ________________ ☐ None

2. Specification of septic tank: ☐ Yes, ____________ ☐ None

3. Desludging of septic tank: ☐ Yes, ____________ ☐ None

4. Disposal at treatment facility: ☐ Yes, ____________ ☐ None

5. Private sector involvement: ☐ Yes, ____________ ☐ None

6. Effluent standard ☐ Yes, ____________ ☐ None

8. Levy ☐ Yes, ____________ ☐ None

**PROCEDURES**

1. Desludging of septic tank: ☐ Yes, ____________ ☐ None

2. Transportation of septage: ☐ Yes, ____________ ☐ None

3. Treatment of septage: ☐ Yes, ____________ ☐ None

4. Customer relations: ☐ Yes, ____________ ☐ None

5. Payment of service: ☐ Yes, ____________ ☐ None

# ANNEX B
# QUESTIONNAIRE FOR SEPTIC TANKS ASSESSMENT

No: _________ Date: __ /__/__

---

**BUILDING INFORMATION**

1. Name: ________________________________________
2. Address: Street ________________________________
   RT/RW ________________________________
   County/Village ________________________________
   Sub-district ________________________________
3. Types of building:
   ☐ Residential ☐ Restaurants
   ☐ Institutional ☐ Educational
   ☐ Business ☐ Hotel
   ☐ Home industry ☐ Social
   ☐ Others: ______________________________
4. Number of users: ____________ Persons
5. Status of building:
   ☐ Private ☐ Official
   ☐ Rental ☐ Others: _________________
6. Number of toilets: _____
7. Number of septic tanks: _____
8. Road access: ☐ Accessible by a truck ☐ Accessible by a small car

☐ Accessible by a motorcycle ☐ No access

9. Water supply and sanitation service
   ☐ water supply, Customer ID: _______________
   ☐ Piped wastewater, Customer ID: _______________

---

| | Submission | Verification | Data Entry |
|---|---|---|---|
| Name: | | | |
| Date: | __/___/____ | __/___/____ | __/___/____ |
| Signature | | | |

## SEPTIC TANK FACILITY

### GENERAL INFORMATION

1. Location: ☐ Off site ☐ On site ☐ Others: _______________________
2. Coordinate point (GPS) ______________________
3. Built in year: ______
4. Last desludging date: ______
5. Distance to closest well: ______ Meter
6. Connection: ☐ Latrine ☐ Bathroom ☐ Rainwater drain ☐ Others: ____________________

### ACCESSIBILITY

7. Distance to access road: ______ Meter
8. Length of hose: ______ Meter

### SPECIFICATION

9. Dimension: $m^3$ (length .... m; width .... m; depth ...m)
10. Type ☐ Fabrication ☐ Non-fabrication

11. Septic tank material: ☐ Stone pair ☐ Reinforced concrete
☐ Concrete ☐ Plastic or fiberglass
☐ Others: ________________________

12. Compartment: ....

13. Access hole: ☐ Yes ☐ No, (explain) ________________

14. Infiltration field ☐ Yes ☐ No, (explain) ________________

15. Ventilation pipe ☐ Yes ☐ No, (explain) ________________

# ANNEX C

# DRAFT OF MAYOR REGULATION ON SEPTAGE MANAGEMENT

**REGULATION OF THE MAYOR OF SURAKARTA**

REGARDING SEPTAGE MANAGEMENT

BY THE GRACE OF GOD ALMIGHTY

MAYOR OF SURAKARTA,

Considering :

a. that in order to protect and improve the condition of sanitation, health and welfare of the people, it is necessary to carry out septage management;
b. that the septage from onsite unit has not been properly managed, causing pollution to the surrounding environment;
c. that septage management is a public service, so it is the regional government's duty to determine policies;
d. that based on considerations as referred to in letter a, letter b and letter c, it is necessary to stipulate a Mayor Regulation regarding Scheduled Septage Management;

In view of :

1. Law No. 16 Year 1950 regarding the Establishment of Big Cities in the Provinces of East Java, Central Java, West Java and in the Special Region of Yogyakarta (Statute Book of the Republic of Indonesia Year 1950 No. 45);
2. Law Number 5 Year 1962 regarding Regional Enterprises (State Gazette of the Republic of Indonesia 1962 Number 10, Supplement to the State Gazette of the Republic of Indonesia Number 2387);
3. Law Number 32 Year 2004 regarding Regional Government (Statute Book of the Republic of Indonesia Number 125 Year 2004, Supplement to Statute Book of the Republic of Indonesia Number 4437) as amended several times, most recently by Law Number 12 Year 2008 regarding the Second Amendment to the Law Number 32 Year 2004 regarding Regional Government (State Gazette of the Republic of Indonesia Year 2008 Number 59, Supplement to the State Gazette of the Republic of Indonesia Number 4844);
4. Law Number 32 Year 2009 regarding Environmental Protection and Management (State Gazette of the Republic of Indonesia Year 2009 Number 140, Supplement to the State Gazette of the Republic of Indonesia Number 5059);

5. Law Number 12 Year 2011 regarding Formation of Laws and Regulations (State Gazette of the Republic of Indonesia Year 2011 Number 82, Supplement to the State Gazette of the Republic of Indonesia Number 5234);
6. Government Regulation Number 82 Year 2001 regarding Management of Water Quality and Water Pollution Control (State Gazette of the Republic of Indonesia Year 2011 Number 153, Supplement to the State Gazette of the Republic of Indonesia Number 4161);
7. Government Regulation Number 16 Year 2005 regarding Development of Water Supply System (State Gazette of the Republic of Indonesia Year 2005 Number 33, Supplement to State Gazette of the Republic of Indonesia Number 4490);
8. Regional Regulation of Surakarta Number 3 Year 1999 regarding Wastewater Management (Surakarta City Regional Gazette Number 17 Year 1999 Series: D Number: 2);
9. Regional Regulation of Surakarta Number 3 Year 1977 regarding Establishment of Surakarta Drinking Water Company (Surakarta Municipality Sheet Number 3 Year 1977 Series: D Number: 2) as amended by the City Regulation of Surakarta Number 1 Year 2004 regarding Amendment to Regional Regulation of Surakarta Number 3 Year 1977 regarding Establishment of Surakarta Drinking Water Company (Surakarta City Regional Gazette Number 1 Year 2004 Series: E Number: 1);
10. Surakarta City Regional Regulation Number 2 Year 2006 regarding Environmental Control (Surakarta City Regional Gazette Number 13 Year 2010);
11. Regional Regulation of the City of Surakarta Number 8 Year 2005 regarding Buildings (Municipal Gazette of the City of Surakarta Year 2009 Number 9).

HAS DECIDED:

To Establish : THE REGULATION OF SURAKARTA MAYOR REGARDING SEPTAGE MANAGEMENT

CHAPTER I
GENERAL PROVISIONS

Article 1

In this Regulation of Surakarta Mayor:

1. Area refers to the city of Surakarta.
2. City Government refers to the Mayor and City Apparatus as an element of the city administration.
3. The Mayor refers to the Mayor of Surakarta City.
4. The Environmental Agency (Badan Lingkungan Hidup), hereinafter abbreviated as BLH, refers to the Surakarta City Environmental Agency.
5. City Office of Public Works (Dinas Pekerjaan Umum), hereinafter abbreviated as DPU, refers to the Surakarta City Office of Public Works.
6. City Office of Health (Dinas Kesehatan Kota) hereinafter abbreviated as DKK, refers to the Surakarta City Office of Health.

7. City Office of Transportation, Communication and Information (Dinas Perhubungan, Komunikasi dan Informatika), hereinafter abbreviated as DISHUBKOMINFO, refers to the Surakarta City Office of Transportation, Communication and Information.
8. City Office of Spatial Planning (Dinas Tata Ruang Kota), hereinafter abbreviated as DTRK, refers to the Surakarta City Office of Spatial Planning.
9. Regional Drinking Water Company (Perusahaan Daerah Air Minum), hereinafter abbreviated as PDAM, refers to the Surakarta City Regional Drinking Water Company.
10. The person in charge of the building refers to the building owner or non-building owner based on an agreement with the building owner, who is responsible for the functioning of the local unit.
11. Domestic wastewater, hereinafter referred to as wastewater, is wastewater generated from household activities, either graywater or blackwater.
12. Local wastewater treatment unit, hereinafter referred to as local unit, is a watertight tank structure that is designed and used to collect and/or treat wastewater located onsite, such as septic tanks.
13. Septage refers to a mixture of human excreta, water and solid wastes that are disposed of in onsite wastewater system.
14. Treated septage refers to biosolid waste resulting from the treatment of domestic sewage in a treatment facility.
15. Septage management is a comprehensive and integrated systematic effort undertaken by the city government to manage septage in a treatment facility from collecting, transporting, processing to utilizing treated septage.
16. Septage desludging refers to the procedure of emptying septic tank by desludging.
17. Mandatory desludging, hereinafter referred as compulsory desludging, is an effort to desludge septage periodically which is required by the city government for every local unit in its area.
18. Septage collection refers to a controlled effort to transport septage to the treatment facility.
19. Scheduled septage desludging refers a periodic septage desludging service.
20. Building Construction Permit (Izin Mendirikan Bangunan), hereinafter abbreviated as IMB, refers to a permit granted by the Mayor to the building owner to build new, change, expand, reduce and/or maintain the building in accordance with applicable administrative and technical requirements.
21. Wastewater Treatment Facility (Instalasi Pengelolaan Air Limbah), hereinafter referred as IPAL, is a technical building and its equipment used to treat wastewater to meet the specific wastewater quality standards.
22. Septage Treatment Facility, hereinafter referred to as STP, is a technical building and its equipment used to treat sludge to meet the specific quality standards for wastewater and processed sludge.

## CHAPTER 2
## PRINCIPLE, GOAL AND OBJECTIVES

### Article 2

1. Septage management is carried out with the principles of responsibility, social benefit, justice and sustainability.
2. Septage management is implemented as part of the urban domestic wastewater management system.
3. Septage management aims to:
    a. Establish a complete, efficient, effective and environmentally friendly urban domestic wastewater treatment system.
    b. Protect the city against environmental pollution and public health diseases.
    c. Maintain the minimum quality standards and function of environment –particularly related to water and soil resources.

   d. Realize an active community participation in maintaining environmental sanitation.
   e. Increase potential sources of local own revenue.
4. The objectives of septage management in this regulation are as follows:
   a. Onsite septic tank systems will have to be in compliance with requirements and registered;
   b. Desludging of septic tanks must be carried out periodically;
   c. Septage transportation must be monitored to avoid environmental pollution;
   d. Treatment of septage is carried out at a specific treatment facility;
   e. Safe use of treated septage.

## CHAPTER 3
## SEPTAGE MANAGEMENT

### Part One
### Onsite Unit Control

#### Article 3

1. All building is required to have access to safe wastewater management services, either a connection to urban septage management system or onsite system that in compliance with technical and administrative requirements.
2. The technical provisions of the onsite unit are as follows:
   a. Location:
      - Have a safe distance from a shallow ground water;
      - Reachable by desludging equipment;
      - No permanent structures should be built over the unit;
   b. Materials being used should be waterproof, strong and leakproof;
   c. Has an inlet, an outlet and a vent pipe;
   d. Has a manhole that is equipped with a lid that can be opened;
   e. Other technical provisions stipulated by DPU.
3. The administrative provisions of the onsite unit are as follows:
   a. Have a building construction permit as applied to either construction of the main building or specifically onsite unit;
   b. Have a legal entity which is responsible for maintaining high functional and performance of onsite unit;
   c. Have a registration number issued by the PDAM.
4. In the next 5 years since this Regulation of Mayor is enacted, all onsite units within the city of Surakarta must meet technical and administrative requirements.

### Part Two
### Septage Desludging

#### Article 4

1. Registered onsite unit must be desludged at least once in every 2–5 years.
2. The local authority responsible for desludging of onsite units is PDAM.
3. The tariff of onsite desludging service shall be determined by the Mayor as calculated and proposed by PDAM.
4. Mandatory desludging should be carried out by septage trucks or other motorized vehicles that are subject to the following conditions:

a. Comply with technical and roadworthiness requirements;
b. Has a vacuum pump, suction hose and other sludge handling equipment;
c. Has a strongly installed tank that is waterproof, rustproof and leakproof;
d. And other requirements determined by DISHUBKOMINFO.

Part Three
Septage Transportation

Article 5

1. Septage is transported to a specific treatment facility.
2. Septage should be safely transported to prevent any form of adverse environmental impacts.
3. Septage transportation must provide a recording system stating information on (a) location and registration number of desludging, (b) operating officer of desludging, (c) desludging time, (d) volume of desludging, (e) location of treatment facility and (f) drop-off time.
4. Septage transportation should be carried out by septage trucks or other motorized vehicles that are subject to the following conditions:
    a. Comply with technical and roadworthiness requirements;
    b. Has a vacuum pump, suction hose and other sludge handling equipment;
    c. Has a strongly installed tank that is waterproof, rustproof and leakproof;
    d. And other requirements determined by DISHUBKOMINFO.

Part Four
Septage Treatment

Article 6

1. Treatment of septage is carried out at a specific treatment facility managed by PDAM.
2. Septage must be treated to produce effluents that meet quality standards in accordance with the provisions of the applicable laws and regulations.
3. BLH is obliged to carry out regular monitoring and evaluation of the septage treatment facilities performance.
4. The treated septage will be used as fertilizers.

CHAPTER 4
OPERATOR

Article 7

1. PDAM is responsible as septage management operator.
2. PDAM is required to:
    a. Develop a plan and procedure for onsite unit registration;
    b. Plan and organize scheduled septage desludging services;
    c. Develop and run standard operating procedures that include both aspects of technical and non-technical (customer, operational, and financial);
    d. Develop and implement a septage management information system;
    e. Plan and implement a control system for septage transportation;
    f. Promote septage management;

    g. Implement the occupational safety and health principles for workers at any step of the operation, from desludging, transporting to treatment of septage;
    h. Report to Mayor on the implementation of septage management.
3. PDAM, in providing desludging and/or transportation services, can cooperate with other parties, including private sectors.
4. Contracted private service providers must fulfill the following requirements: (a) have legal status, (b) have a business license, (c) have trained personnel, (d) have equipment and fleets in accordance with technical provisions, (e) other requirements determined by the PDAM.

## CHAPTER 5
## SUPERVISION

### Article 8

1. The implementation of septage management system is under the supervision of the City Government.
2. Supervision as referred to in Article 8 Sub-article 1 is carried out through regular monitoring and evaluation at least once in every 6 (six) months.

## CHAPTER 6
## CONCLUSION

### Article 9

1. The Regulation of Mayor becomes effective on the date of its establishment.
2. For public cognizance, this Regulation of Mayor shall be promulgated by placing it in the Surakarta City Gazette.

Issued in Surakarta
MAYOR OF SURAKARTA,

# ANNEX D

# PRIVATE SECTOR PARTNERSHIP AGREEMENT FOR SCHEDULED DESLUDGING

**COOPERATION AGREEMENT**

**BETWEEN**

**THE WASTEWATER UTILITY**

**WITH**

**COMPANY. ________________,**

**ON SCHEDULED DESLUDGING SERVICES**

Today ______ on (date) ______ (month)_____ (year)______________________________ ( __ /__ /____ ), the undersigned below:

1. OPERATOR **.....** hereinafter referred as the First Party.

2. COMPANY .... hereinafter referred as the Second Party.

The First and the Second Party, together hereinafter referred to as "PARTIES" and/or individually hereinafter referred to as "PARTY".

THE PARTIES hereby first explain:

1. That THE FIRST PARTY is a Regional Company which is responsible for the management of sludge management in the city of ..... as referred to in the Mayor Regulation number ......
2. That THE FIRST PARTY requires a pioneering process to ensure Scheduled Desludging Services as a major component of the sludge management system can be better implemented in all areas of the city of ......
3. That THE FIRST PARTY requires the involvement of the SECOND PARTY as a desludging service provider that is already operating in the city of ...... and is willing to work together to improve the technical aspects of the operation of Scheduled Desludging Services.

Based on the foregoing matters, the PARTIES agree to enter into a cooperation agreement in the implementation of Scheduled Desludging Services at the stage of … in accordance with the following terms and conditions:

**Article 1**
General Requirements

1. Septage is a mixture of solids and liquids, including human waste that has accumulated in the local onsite domestic wastewater unit.
2. Septage management is a comprehensive and integrated systematic effort undertaken by the regional government to manage the adverse impact of septage by controlling the onsite units, desludging, transporting, processing and utilizing treated sludge.
3. Local treatment unit, hereinafter referred to as onsite unit, is a watertight tank facility that is designed and used to accommodate and/or treat wastewater located in one land/location with its wastewater sources, such as septic tanks and such.
4. Treated sludge is solids which is the result of the processing of sewage sludge at a treatment plant facility.
5. Desludging is a procedure for excreting sludge from an onsite unit using mechanical equipment.
6. Septage transportation is a controlled effort to transport sludge to the specified treatment plant facility (Wastewater Treatment Plant).
7. Septage disposal is a procedure for the reduction of septage from its transport vehicle at the receiving unit in specified treatment plant facility (Wastewater Treatment Plant) in which location is previously determined and agreed upon.
8. Mandatory desludging, hereinafter referred to as compulsory desludging, is an effort to collect desludging periodically which is obliged by the city government to every onsite unit in its area.
9. Scheduled Desludging Services are services provided to support the compulsory desludging of each onsite unit throughout the city area and followed by transporting and disposing sludge at the treatment plant facility (Wastewater Treatment Plant).
10. The customer of scheduled desludging services, hereinafter referred to as a person or a registered entity who is responsible for the functioning of the onsite unit in a building and from receiving scheduled desludging services.
11. Wastewater Treatment Plant, hereinafter referred to as WWTP, is a technical building and its equipment used to treat wastewater until it meets the specified wastewater quality standards.
12. Septage Treatment Plant, hereinafter abbreviated as STP, is a technical building and its equipment used to treat sludge until it meets the specified quality standards for wastewater and treated sludge.
13. Scheduled Desludging Management Information System hereinafter referred to as MIS is software that is prepared to assist in arranging desludging schedules and recording desludging performance within a certain period of time as well as calculating desludging service compensation and facilitating information exchange between the two parties using Android-based technology.
14. Desludging vehicles are trucks or other motorized vehicles that fulfill the requirements for desludging and/or transportation of septage.
15. Scheduled Desludging Work Order is an assignment order issued by the FIRST PARTY to the SECOND PARTY to provide scheduled desludging services to a group of customers within the timeframe specified in the Order.

16. Stage .............. is the stage where the FIRST PARTY organizes Scheduled Desludging Services on a limited basis, especially only to groups of residential and/or residential customers within the scope of limited service areas. In stages … , FIRST PARTY is expected to be able to gain direct experiences from field operations which will later be used to improve management and operation procedures for Scheduled Desludging Services.
17. Scheduled desludging trip plan, hereinafter referred to a series of trips from a stool sludge vehicle to transport sludge in one or more buildings followed by transportation of sludge to the specified WWTP/STP.

**Article 2**
Purpose and Objectives

1. THE FIRST PARTY collaborates with the SECOND PARTY to provide Scheduled Desludging Services during the time span of the Stage .............. in several designated areas of the city
2. The SECOND PARTY provides input to the FIRST PARTY to improve the technical and administrative procedures of Scheduled Desludging Services based on experience gained during the implementation stage.

**Article 3**
Scope of Cooperation

1. Conducting desludging, transportation and disposing sludge operations are part of the implementation of Scheduled Desludging Services;
2. An assessment of the condition and performance of is conducted at the onsite units;
3. Implementation of administrative aspects of Scheduled Desludging Services;
4. Completion of technical and administrative procedures for Scheduled Desludging Services.

**Article 4**
Period of time

1. This Agreement is valid since the Stage .............. of the operation of Scheduled Desludging Services which is at least 6 (six) months from the signing of this Agreement;
2. This agreement can be extended and/or improved according to the needs and agreements of both PARTIES which will then be set forth in the form of an Amendment to the Agreement signed by the PARTIES.

**Article 5**
Rights and obligations

1 FIRST PARTY has the rights to:
  a. Enact Operational Guidelines for the Implementation of Scheduled Desludging Services which must be carried out by the SECOND PARTY as set out in the Annex to this Cooperation Agreement;
  b. Determine desludging vehicles and operating officers that can be hired by the SECOND PARTY in the operation of Scheduled Desludging Services by referring to the terms and technical specifications listed in the Appendix to this Cooperation Agreement;

c. Determine the SECOND PARTY working area in organizing Scheduled Desludging Services as stated in the Annex to this Cooperation Agreement;
d. Determine the location and schedule for the implementation of desludging operations which must be carried out by the SECOND PARTY as stated in each order;
e. Determine WWTP and/or STP where the SECOND PARTY must carry out sludge as stated in each order;
f. Request the SECOND PARTY to repair or replace the sludge vehicles which are deemed to no longer meet the requirements and specifications as contained in the Appendix to this Cooperation Agreement;
g. Order the SECOND PARTY to suspend or cancel desludging and/or transportation and/or disposal of sewage sludge;
h. Determine the amount of compensation money for the SECOND PARTY services by referring to the Directors Decree .............. and listed in the Appendix to this Cooperation Agreement;
i. Monitor and evaluate the performance of the SECOND PARTY in implementing Scheduled Desludging Services;
j. Give sanctions to the SECOND PARTY in accordance with the provisions stipulated in this Cooperation Agreement.

2 FIRST PARTY has the obligations to:
a. Issue certificates required by the SECOND PARTY in the licensing process for the operation of sludge vehicles which will be used in the operation of Scheduled Desludging Services;
b. Provide technical and administrative training to operating personnel assigned to the SECOND PARTY in the operation of Scheduled Desludging Services;
c. Issue identification of Scheduled Desludging Services for sludge vehicles and SECOND PARTY operating personnel who have been assessed as fulfilling the requirements and specifications listed in the Appendix to this Cooperation Agreement;
d. Issue an order to the SECOND PARTY during the period of the cooperation agreement and according to the minimum number of reservations as stated in the Attachment to this Cooperation Agreement;
e. Determine WWTP and STP to be used in the implementation of Scheduled Desludging Services as stated in the Annex to this Cooperation Agreement;
f. Ensure that the WWTP and STP are in good operating condition and can be reached by the SECOND PARTY desludging vehicles;
g. Receive and process sludge transported by the SECOND PARTY according to the procedure stated in the Appendix to this Cooperation Agreement;
h. Prohibit desludging vehicles that do not yet have identification to carry out Scheduled Desludging Services;
i. Pay compensation in accordance with the performance of the SECOND PARTY services rendered in the amount as stated in the Appendix to this Cooperation Agreement.

1 SECOND PARTY has the rights to:
a. Obtain identification of Scheduled Desludging Services from the FIRST PARTY for desludging vehicles and operating personnel who have been assessed as meeting the requirements and specifications listed in the Appendix to this Cooperation Agreement;
b. Receive an order from FIRST PARTY for the duration of the cooperation agreement and in accordance with the minimum number of reservations as stated in the Attachment to this Cooperation Agreement;

c. Determine the location and daily schedule for the implementation of compulsory suction operations by referring to the order given by the FIRST PARTY;
d. Disposing sludge at a specified WWTP and/or STP determined by the FIRST PARTY as stated in the Attachment to this Cooperation Agreement Letter;
e. Obtain payment of Compensation Services from the FIRST PARTY in accordance with the performance of the SECOND PARTY services with the amount as stated in the Appendix to this Cooperation Agreement.

2 SECOND PARTY has the obligations to:

a. Have permits related to business continuity and feasibility of sludge vehicles issued by the competent authority to support the operation of desludging and/or transportation and/or disposal of sludge;
b. Run the Operational Directive on the Implementation of Scheduled Desludging Services issued by the FIRST PARTY as stated in the Appendix to this Cooperation Agreement;
c. Use desludging vehicles and operating personnel who already have ID scheduled desludging operator;
d. Provide Scheduled Desludging Services according to location and implementation schedule as stated in the order issued by the FIRST PARTY;
e. Dispose sludge at a specified WWTP and / or STP determined by the FIRST PARTY as stated in each order;
f. Carry out administration aspects of Scheduled Desludging Services determined by the FIRST PARTY;
g. Have a communication device to support the use of MIS;
h. Fulfill the orders of the FIRST PARTY, specifically related to the repair or replacement of desludging vehicles and related to the delay or cancellation of desludging and/or transportation and/or disposal of sludge;
i. Assist the FIRST PARTY in assessing the condition and performance of onsite units;
j. Provide input and recommendations to the FIRST PARTY regarding various matters to improve technical and administrative procedures for Scheduled Desludging Services;
k. Receive sanctions given by the FIRST PARTY in accordance with the provisions stipulated in this Cooperation Agreement;
l. Return the ID of the Scheduled Desludging Service to the FIRST PARTY at the time this cooperation agreement ends.

**Article 6**

Costs

1. Operational Performance Guarantee; THE FIRST PARTY does not charge the SECOND PARTY Operational Performance Guarantee during the Stage .............. of Scheduled Desludging Services.
2. Service Compensation; THE FIRST PARTY will pay Service Compensation in the amount of ...... to the SECOND PARTY for each Scheduled Desludging Service.
3. The amount of compensation for services will be evaluated annually and adjustments can be considered due to the following conditions and the general economic situation.

**Article 7**

Payment method

1. Service Compensation Money as referred to in Article 5 paragraph (2) is paid by the FIRST PARTY to the SECOND PARTY 1 (one) month no later than the 10th (tenth) of the following month.
2. The amount of compensation as referred to in paragraph (1) will be paid in cash through the FIRST PARTY General Cash.
3. The SECOND PARTY will submit a billing receipt with the accompanying supporting documents as a condition for the FIRST PARTY to make payments.
4. Supporting documents as referred to in paragraph (3) include:
   a. Order;
   b. Report on Work Results that have been signed by relevant parties (customers, WWTP/STP officers and Head of Liquid Waste Sector);
   c. Print a Record of Execution of Works issued by the Scheduled Desludging Services Information System.

**Article 8**

Security and Safety

1. The PARTIES are responsible for all the work completeness for their operating personnel to support security and work safety in organizing Scheduled Desludging Services.
2. THE PARTIES agree to carry out the safety and work safety provisions issued by the FIRST PARTY.
3. The SECOND PARTY provides insurance for its operating operators involved in the operation of Scheduled Desludging Services.

**Article 9**

Prohibition

1. THE FIRST PARTY is prohibited from collecting any fees to the SECOND PARTY, whether directly or indirectly related to the operation of Scheduled Desludging Services.
2. SECOND PARTY is prohibited from disposing of sludge other than in the designated WWTP and/or STP.
3. The SECOND PARTY discharges other wastewater other than sludge at the designated WWTP and/or STP.

**Article 10**

Transfer of Rights and Obligations

The PARTIES are not allowed for any reason to transfer some or all of their rights and obligations as stated in this Cooperation Agreement to other parties, without the written approval of the PARTIES.

**Article 11**

Supervision and Control of Work Implementation

1. THE FIRST PARTY assigns the Head of the Liquid Waste Sector, i.e. The Head of the Liquid Waste Management Installation Section to supervise and control the implementation of this cooperation agreement;

2. THE FIRST PARTY uses MIS to assist in the regulation, control and supervision of the implementation of Scheduled Desludging Services by the SECOND PARTY in an integrated manner;
3. THE FIRST PARTY may at any time carry out unannounced inspections and carry out random surveys of the implementation of the SECOND PARTY's work, including checking the characteristics of the sludge carried by the SECOND PARTY to the WWTP and/or STP.

**Article 12**

Legal Guarantee

1. THE FIRST PARTY hereby declares to carry out the cooperation referred to in this Agreement in a professional manner and guarantees and frees the SECOND PARTY from any interference from other parties in carrying out this work;
2. The SECOND PARTY hereby declares to carry out cooperation as referred to in this Agreement as professionally as possible and guarantees and exempts the FIRST PARTY from all customer complaints in connection with the work of the SECOND PARTY as referred to in this Agreement;
3. In the event of a customer complaint related to the work of the SECOND PARTY, the complaint will be forwarded to the SECOND PARTY for immediate action.

**Article 13**

Termination of Agreement

1. This agreement can be ended if there is an agreement between THE PARTIES, even though the period referred to in Article 4 paragraph (1) has not yet ended;
2. In the event that one of the PARTIES intends to terminate this Agreement before the end of the term of this Agreement, the party who will terminate it must first notify in writing to the other party no later than 15 (fifteen) Business Days prior to the date proposed for termination of this Agreement;
3. Each PARTY may terminate this Agreement unilaterally in the event that the obligations of the other party are not carried out as stipulated in this Agreement, after 3 (three) consecutive summons (written warning) have been carried out;
4. This agreement may be terminated or cancelled automatically, if there are provisions of the laws and/or Government policies that do not allow this Agreement to take place;
5. In the event of termination/termination of this Agreement, the PARTIES agree to waive the provisions of Article 1266 of the Civil Code;
6. The termination of this Agreement does not reduce and do not waive the rights and obligations of each PARTY that arises before the termination of this Agreement;
7. In the event that the termination of this Agreement is carried out due to negligence of one of the PARTIES, all claims and requests for compensation from third parties to the injured Party arising from the termination of this Agreement shall be the responsibility of the Party who committed negligence.

**Article 14**

Declaration of Guarantee of the Parties

1. THE PARTIES hereby declare to ensure that the parties who sign this Agreement are legal and authorized in accordance with the provisions of the Articles of Association of the company they

represent and because the PARTIES release each other over the demands of other parties regarding the signing of this Agreement;

2. This agreement does not contradict the Articles of Association of each party and does not violate Government regulations that must be obeyed by each PARTY in running the company;
3. If there is an article or paragraph from this Agreement which is declared null and void by law or a legal defect by the Court, then it will not affect the validity or validity of the verses and/or other articles in this Agreement, so that the provisions others in this Agreement remain valid and binding;
4. The PARTIES hereby declare that the enforcement of the clauses governing technical cooperation in services as regulated in this Agreement will apply in accordance with the type of service and the stage of the collaboration carried out.

**Article 15**
Force Majeure

1. Force Majeure is any situation or event that occurs outside the authority of the PARTIES, including but not limited to riots, epidemics, fires, floods, earthquakes, general strikes, war, government decisions that are directly and materially prohibit the PARTIES from directly carrying out their obligations in accordance with this Agreement;
2. In the event of one or several incidents and/or events as referred to in paragraph (1) of this Article, the PARTY affected by Force Majeure is obliged to notify the other PARTY in writing no later than 14 (fourteen) Calendar Days as of the occurrence of the event;
3. The events referred to in paragraph (1) may be made as a prolongation of the implementation of obligations by the PARTY who experiences Force Majeure and therefore frees the PARTY who experiences such Force Majeure and sanctions for holding out in carrying out the obligations that should be fulfilled;
4. All losses and costs suffered by the PARTIES as a result of the occurrence of Force Majeure are the responsibility of each PARTY;
5. If the Force Majeure lasts more than 3 (three) months, then one of the PARTIES can terminate this Agreement by notifying in writing that the PARTY that has experienced Force Majeure and this Agreement will expire since receipt of the said notification by the party experiencing Force Majeure and approved by Other PARTIES.

**Article 16**
Change

Any amendments and matters that have not been regulated or have not been sufficiently regulated in this Agreement as a step to improve the implementation of this Agreement will be determined later by deliberation by the PARTIES and will be set forth in the Amendment to the Agreement (Addendum) which is an integral and inseparable part of this Agreement.

**Article 17**
Dispute resolution

1. The validity, interpretation, and implementation of this Agreement shall be fully regulated and subject to legal regulations in force in the Republic of Indonesia.

2. Disputes arising between PARTIES in the implementation of this Agreement will be resolved by deliberation to reach consensus.
3. If the deliberation method to reach consensus is not reached, the PARTIES agree to resolve all disputes arising through ……

**Article 18**

Attachment

1. This Agreement is accompanied by attachments which are an integral and inseparable part of this Agreement and have binding legal force as is the case with this Agreement.
2. Attachments as referred to in paragraph (1) include:
   a. The scope of Joint Operations, at least mentions (i) working area, (ii) time span, (iii) disposal location and (iv) compensation for services;
   b. Scheduled Desludging Service Operations;
   c. Terms and Technical Specifications for Desludging Vehicles;
3. Changes to the contents of this Agreement that change, add or delete the contents of this Agreement and or other matters that have not been regulated in this Agreement are made an Additional Agreement (Addendum) which is a binding and inseparable part of this Agreement.

**Article 19**

Closing

This Agreement was made and signed by THE PARTIES on the day, month and year mentioned in 2 (two) original copies, each stamped with sufficient stamp duty and has the same legal force for the PARTIES, 1 (one) duplicate for the FIRST PARTY and 1 (one) copy for the SECOND PARTY.

| **SECOND PARTY,** | **FIRST PARTY,** |
|---|---|
| COMPANY ……….. | THE CITY OF … SCHEDULED DESLUDGING OPERATOR …………… |
| Director | President Director |

# ANNEX E
# SOP FOR SEPTIC TANK DESLUDGING

**STANDARD OPERATING PROCEDURE (SOP)**
**Desludging of Septic Tank**

| | |
|---|---|
| **Government of Surakarta** | **PDAM Surakarta** |
| **SOP Number:** | |
| **SOP Name:** | **Desludging of Septic Tank** |
| **Date of Manufacture:** | **10 August 2015** |
| **Revision Date:** | **4 November 2015** |
| **Effective date:** | |
| **Endorsed by:** | **Directors of<br>Regional Drinking Water Company<br>Kota Surakarta<br>President Director** |

<table>
<tr><th>LEGAL BASIS:</th><th>USER:</th><th>QUALIFICATION TO PERFORM:</th></tr>
<tr><td><ul><li>Surakarta Mayor Regulation No. 16-A 2014 - Management of sewerage</li></ul></td><td><ul><li>Desludging partner manager</li><li>Desludging crew member</li></ul></td><td><ul><li>Physically and mentally healthy</li><li>Understand the rules and regulations concerning scheduled desludging</li><li>Able to explain the rights and obligations of the customer</li><li>Able to operate MIS Mobile</li></ul></td></tr>
<tr><th>LINKAGES:</th><th colspan="2">EQUIPMENT SUPPLIES:</th></tr>
<tr><td><ul><li>SOP for Desludging Preparation</li><li>SOP for Septage Transportation</li><li>SOP for Septage Disposal</li></ul></td><td><ul><li>Desludging truck</li><li>Road sign during operation</li><li>Desludging hose</li><li>Depth measuring stick</li><li>Flashlights</li></ul></td><td><ul><li>Crowbar and equipment</li><li>Bucket, cleaning cloth and equipment</li><li>Personal protective equipment, (rubber boots, overalls, gloves, goggles and cap).</li><li>Smartphone.</li></ul></td></tr>
<tr><th>WARNING:</th><th colspan="2">REGISTRATION AND DATA COLLECTION:</th></tr>
<tr><td><ul><li>Desludging operation can only be performed by an official partner.</li><li>Only individual with formal Letter of Assignment can perform desludging</li><li>Only individual with a driver license can operate the desludging truck.</li><li>The truck must follow the specified route.</li></ul></td><td colspan="2"><ul><li>Desludging Sheet: A sheet used by the crew member to record the implementation of desludging operation in each building, containing identity of customer, customer's address, identity of desludging partner, information of the septic tank, date and time of operation, boxes of signatures from the crew member and the customer.</li><li>Transportation Control Card: A card used by the crew member to record the septage disposal in the wastewater / septage treatment plant, containing the name of the treatment plant, volume of disposed septage, date and time of operation, boxes of signatures from the crew member and the treatment plant officer.</li><li>Septic Tank Inspection Form.</li></ul></td></tr>
</table>

# FLOW DIAGRAM OF THE PROCEDURE FOR DESLUDGING OF SEPTIC TANK

| | Activities | Crew 1 | Crew 2 | Customer | | Materials | Time | output |
|---|---|---|---|---|---|---|---|---|
| 1 | Desludging truck moves as planned and through the agreed route to destination point. | | | | | Desludging Sheet | 10-30 mins | Truck moves |
| 2 | Desludging truck is parked in a safe place and do not interfere with the surrounding traffic. | | | | | Desludging Sheet | 5 mins | Arrive in destination |
| 3 | Please note the time of arrival in the Desludging Sheet. | | | | | Desludging Sheet | 5 mins | Time recorded |
| 4 | Meet the homeowner or the representative and present ID card and the scheduled desludging assignment letter. | | | | | Identi-bag Letter Yourself<br>Desludging Sheet | 5 mins | The crew is recognized |
| 5 | Explain the purpose of the visit, the rights and obligations of customers (including the volume of septage to be pumped out) as well as the expected cooperation from the customer (including clean water). | | | | | Desludging Sheet | 15 mins | Customer understand the purpose of the visit |
| 6 | Obtain customer consent for the work,<br>• If agreed, continue.<br>• If not agreed, step 28. | | | yes | no 28 | - | 5 mins | Customers ready to cooperate |
| 7 | Scan building's barcode of the customer. | | | | | Smartphone | 2 mins | Barcode is scanned |
| 8 | Go to the septic tank and make sure the tank lid in an open condition,<br>• If it is open, continue.<br>• If not open, ask the customer to open it. | | yes | no | | - | 5-30 mins | Septic tank's lid is open |
| 9 | Customers open the lid of the tank,<br>• If open, proceed.<br>• If cannot open, step 26. | | | yes | no 26 | - | | Septic tank's lid is open |
| 10 | Use of personal protective equipment (PPE) ie rubber boots, goggles, gloves and hats. | | | | | PPE | 5 mins | PPE used |
| 11 | Check the contents of the septic tank depth and the thickness of the sediment at the bottom of the tank. Use the depth measuring stick and a flashlight. | | | | | Flashlight<br>Measuring stick | 5 mins | Contents of septic tank is known |
| 12 | Determine the path of the desludging hose. | | | | | - | 5 mins | Hose path is specified |
| 13 | Move the desludging truck into a better position for connecting the hose. | | | | | - | 5 mins | Truck in position |
| 14 | Put up signs around the working area implementation of the operation in the vehicle suction feces. | | | | | Work signs | 5 mins | Signs are in place |
| 15 | Connect the suction hose from the desludging and put into the septic tank until it reaches 80% of the depth of the tank. | | | | | Suction hose | 15 mins | Hose is attached |

| | | | | | | | | |
|---|---|---|---|---|---|---|---|---|
| 16 | Estimate the height level of desludging truck tank contents when the desludging volume is reached. | | | | | - | 2 mins | Goal of the tank level is known |
| 17 | Enter data into the MIS mobile in the smartphone | | | | | Smartphone | | |
| 8 | Turning desludging pump and ensure the septage is flowing well. | | | | | - | 2 mins | Pump is on |
| 19 | Monitor the desludged septage,<br>• If the height level is reached, stop the pump.<br>• If the height level has not been reached, continue the pumping. | | | | | - | 5-10 mins | Septage is flowing to the tank |
| 20 | Fill the Septic Tank Inspection Form, including taking a photo (if needed). | | | | | Septic Tank Inspection Form | 10 mins | Filled form |
| 21 | Remove the suction hose after making sure there is no septage left in the hose. | | | | | - | 5 mins | Hose is removed |
| 22 | Clean the suction hose with water in a secure place. | | | | | Clean water<br>Bucket, brush | 5 mins | Clean hose |
| 23 | Roll up the suction hose and restore it to its designated place in the desludging truck. | | | | | - | 5 mins | Hose is back in its place |
| 24 | Clean the surface around the septic tank hole and the location where suction hose are located. | | | | | Clean water<br>Bucket, brush | 5 mins | Clean area |
| 25 | Remove PPE and put it back into its place in the desludging truck. | | | | | PPE | 5 mins | PPE in place |
| 26 | Complete the Desludging Sheet including record the septage volume and time of completion of the work. | | | | | Desludging Sheet | 5 mins | Deslud. Sheet is filled |
| 27 | The desludging Sheet signed by the crew member and customer. | | | | | Desludging Sheet | 5 mins | Deslud. Sheet is signed |
| 28 | Restore all work signs and other items to the place in the desludging truck. | | | | | Sign | 5 mins | Signs are back in place |
| 29 | Enter data into the MIS mobile in the smartphone | | | | | | | |
| 30 | Drive the truck from customer's premises to the wastewater / waste treatment plant or the next customer. | | | | | - | | Truck leaves the premise |

# ANNEX F
# MIS FOR DESLUDGING SERVICE BY IUWASH PLUS PROGRAM

MIS Desludging Service consists of several applications and are designed for special purposes. The IUWASH PLUS program is the pioneer in developing the MIS in Indonesia. The MIS is designed to support wastewater utility in conducting scheduled desludging and on-demand desludging services. The MIS utilizes Geographic Information System (GIS) technology that records the geo-coordinate of customer location, septage treatment plant and desludging trucks. The locations and movements of the trucks are displayed on digital maps so the operations manager can easily know where units are at any given time. The operation of MIS Desludging Service requires Desktop Computer, Data Server, Smartphone, and GPS tracker which is installed in desludging truck. The system also needs a barcode printer which is used to make customer barcode and truck barcode, and certainly the internet connection to allow communication between each device with the data server (see diagram below).

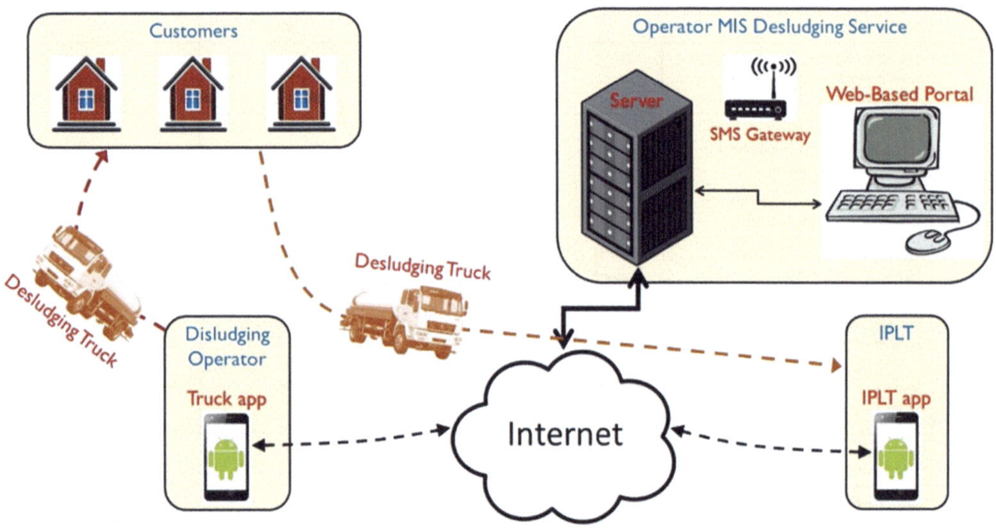


MIS Desludging Service is developed based on the specific needs of a city to run a more efficient, effective and transparent desludging service. Efficient, because the process is done digitally and using internet network which makes communication faster or even real-time. Effective, because it uses the GIS technology to compile a list of customers of the schedule desludging scheme. Transparent, because it can accurately monitor truck movement and septage disposal. The diagram below shows the process flow within the IT-based MIS.

The process flow of desludging operation is as follows:

- Using MIS Desludging Service, the wastewater utility develops database of scheduled desludging customers. Data may come from census, on-demand desludging records as well as from customer application process. Each customer will be given a barcode (customer ID).
- The wastewater utility will require desludging partner to install GPS tracker in each of their trucks along with the barcode. In addition, the crew member must carry a smartphone that has installed a

special application which is connected to the MIS. This will help the utility to record the data of sludge volume and truck movement.

- Another application installed on the Android smartphone is the septage treatment plant (STP) apps for the operator who records the arrival of desludging trucks and the volume of sludge disposed in the treatment plant.
- If the customer database has been developed, the MIS can help the utility to develop monthly desludging work-order which provides the list of customer that will be serviced, schedule of the desludging operation and recommended route for the truck from the desludging location to the treatment plant location. The monthly desludging work-order and its attachments are sent electronically by email to the crew member of desludging truck.

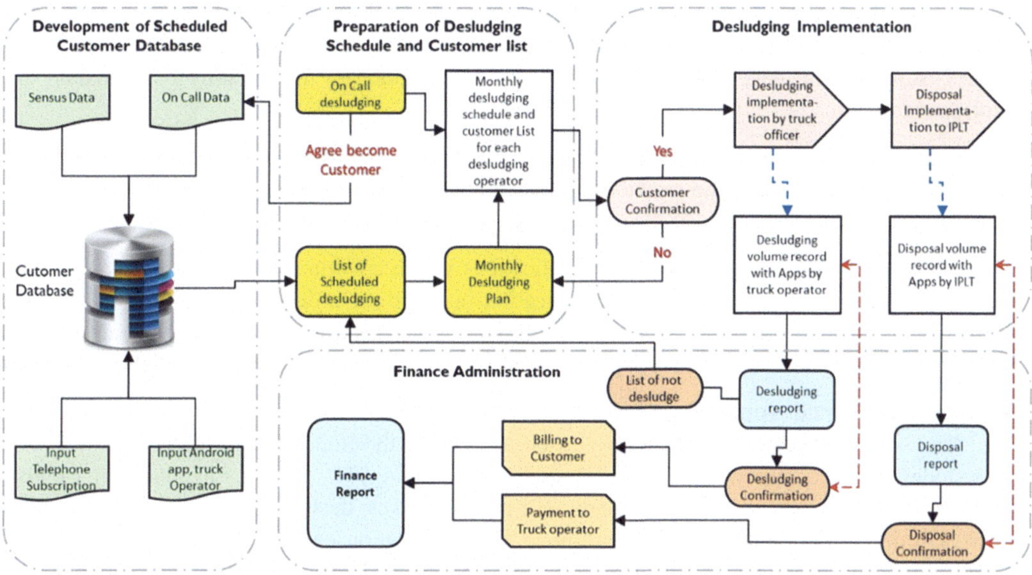


- Upon receiving the monthly work-order, the desludging truck crew will contact the customers in the list to make appointment for the desludging activity. The monthly customer list and the desludging schedule are then detailed into a daily desludging list based on customers' willingness. The daily list is sent to the desludging truck crew as their guidance for operation.
- According to the schedule, the desludging truck will depart from the station to every customer on the list. The crew must use the smartphone apps to record the odometer and sent to the data server.
- Upon reaching the customer's home, the crew must scan the customer's barcode using the smartphone apps to confirm whether the house visited is actually the customer's house listed in the daily order. If it matches, the desludging operation will be conducted. If it does not match, the truck must move and visit the other customer in the list.
- After the desludging is completed, the crew member must record the desludged septage volume by using the truck apps and sent to the data server. This process is repeated to the next customer according to the daily list before the truck moves to the septage treatment plant to dispose the septage.

- By the time the truck reaches the septage treatment plant (STP), the barcode of the desludging truck will be scanned by the treatment plant operator as a proof that the truck has reached the plant site. The volume of disposed sludge is recorded in the STP apps and sent to the data server by the STP operator.
- After the disposal is complete, the truck will return to the station, then the crew will record the odometer/trip meter to the truck smartphone apps and send it to the server.
- Daily desludging process will be repeated in the next day, until the completion of all the list of customers who have confirmed willingness.
- The desludging truck crew reports the implementation of the monthly desludging, informing the septage volume taken from each customer, the mileage of the desludging truck based on daily odometer/trip meter records and list of customers who have not yet received service.
- Data of desludged septage volume and truck mileage sent by the truck apps and data of disposed septage volume sent by STP apps will be processed by the Portal application and will be used to confirm monthly desludging report of a desludging truck operator. The data becomes the basis of the billing to the customer and payment to the truck operator, as well as the financial report.
- The customers who have not been serviced will be included in the monthly list schedule of the following month.

The process of desludging and disposal in real time can be supervised by wastewater utility through Portal desktop application that displays the location and movement of desludging trucks.

**Applications and its functions on MIS Desludging Service**

MIS Desludging Services consists of three applications which are connected to the server data though internet. They are (1) the web-based Portal application installed on a desktop computer, (2) the Truck apps which is installed in the smartphone of the desludging truck crew and (3) the STP apps which is installed in the smartphone of the treatment plant operator. The functions of each application are as follows:

1. **Web-based Portal Application**

   The web-based Portal application is used to manage the implementation process of customer desludging services as well as disposal sludge process in STP. Features available in web-based Portal application such as: development customer databases, desludging truck operator databases, work order system and accounting systems. This application can also be used to monitor truck movement by displaying it with maps on the application dashboard. Functions available in this application are:
   - function of scheduled customer databases management,
   - customer database development stored in the server,
   - customer input data form based on phone registration,
   - customer input data form based on manual census form,
   - new customer data receiving from registration form of truck application,
   - receipt of data from the input form installed in desludging officer smartphone,
   - data import facility from spreadsheet table (xls, csv, dbf, etc.),
   - data import facility of other web-based applications (api service),
   - the creation of customer barcodes based on scheduled customer database,
   - desludging truck operator database,
   - desludging truck operator database development stored in the server,
   - desludging truck database development stored in the server,
   - the creation of desludging truck barcodes based on desludging truck databases,
   - function of STP database management,

- STP database development,
- monitoring the disposal process of each desludging truck,
- acceptance confirmation of the presence of desludging truck in septage treatment plant through barcode scanning,
- acceptance data of disposal volume from every desludging truck at septage treatment plant reported by the officer,
- desludging schedule and customer list preparation,
- preparation of monthly desludging schedule and customer list based on zoning, through; (1) analysis of scheduled customer data, (2) list of desludging requests via telephone, (3) list of remaining customers of previous month,
- preparing of monthly desludging work order for every truck operator based on zoning,
- delivery of work order desludging to the truck operator is equipped with the desludging schedule and the customer list via email,
- desludging process monitoring,
- interactive map to viewing truck location and movement,
- reception of desludging volume data from desludging truck officers,
- reception of daily trucking mileage data from desludging officers,
- reception of sludge disposal volume data from septage treatment plant officers,
- reception report of desludging and sludge disposal implementation from desludging truck operator,
- administrative and financial,
- analysis of desludging volume data, sludge data and truckload mileage data,
- preparation of desludging and disposal process reports,
- preparation of invoice bill to customer,
- preparation voucher payment to truck operator,
- financial statements.

2. **Truck Application**

The Truck application is installed on smartphones android is used to scan customer barcodes to ensure customers are included in the list and desludging schedule. This application will also use to record the volume of sludge has desludged and send the data to the server which is monitored by Portal application. In addition, this application is also given the function of data input of potential customers encountered by the desludging officer in the field, then sent to the server to become customer data. The application also provides a form to record mileage of trip meter and send it to the server. The functions available in this application are:

- downloading desludging schedule and customer list,
- customer confirmation for desludging schedule to prepare daily desludging and customer list,
- daily truckload mileage recording,
- form for odometer/trip meter data recording initial and final,
- delivery record data to server,
- customers barcode scanning,
- desludging process reporting,
- form desludging data recording,
- submission of desludging data to server/portal,
- submission of customer list of monthly unsuccessful desludging to server/portal,
- daily recording workload mileage (odometer/trip meter),

- o record form for truck odometer/trip meter at start point,
- o record form for truck odometer/trip meter at finish point),
- o delivery of daily odometer/trip meter data to server/portal,
- o function of scheduled customer subscription,
- o scheduled customer subscription form,
- o delivery of customer subscription form data to server/portal
- o customer member application/registration,
- o form registration for new customer of scheduled desludging,
- o submission of new customer data to server/portal.

3. **STP Application**

   STP application is installed on the android smartphone of the treatment plant operator to scan truck barcodes, aiming to ensure truck presence at STP locations for sludge disposal. This application also has a function to record disposed septage volume and to send the record to the server. Functions available in this application are:

   - o desludging truck barcode scanning,
   - o septage disposal reporting,
   - o recording form for desludging volume,
   - o submission of disposed septage data to server/portal.

# ANNEX G
# SAFETY ISSUES OF SCHEDULED DESLUDGING

Every service provider of scheduled desludging has a legal duty to ensure the health and safety of the crew members and other workers in the field. They also have the duty to ensure that their desludging truck, equipment, treatment plant and other workplace are compliant with the Occupational Health and Safety provisions. Ensuring occupational health and safety compliance should not be seen as a regulatory burden. On the contrary, it offers significant opportunities, such as reducing costs, risk, employee absence, turnover rates, work accidents, threat of legal action, enhancing reputation among service providers and partners, better reputation for corporate responsibility in the eyes of investors, customers, and the public, and also increasing productivity, because employees are healthier, happier, and more motivated.

## CREATE A SAFETY CULTURE

In the provision and maintenance of scheduled desludging services, crew members and other workers are exposed to serious occupational and environmental health hazards risking illness, injury and death. To prevent them from those consequences, the scheduled desludging service provider should create a strong safety culture and establish a comprehensive safety program.

A safety culture is a broad, organization-wide approach to safety management. It starts with the collection of value and beliefs that employers and employees share in relation to risks in the workplace. An effective way to create a safety culture is leading by example, by following all safety policies and encouraging workers to do the same. If a service provider commits to safety, employees will follow suit. Additionally, workers also need to be trained on a regular basis. Trained workers also embrace safety culture more readily because they are aware of hazards and the effect that they can have from the provision and maintenance of scheduled desludging service.

## POTENTIAL RISKS

The following are harmful risks that potentially can occur in the provision and maintenance of septage management:

- Septage contains hazardous substances such as bacteria, viruses, and worms that can cause serious and life-threatening diseases if not handled properly.
- Improper use of desludging trucks and sludge treatment equipment can cause injury or death.
- Improper design and use of high-voltage electrical equipment can cause electric shock, electrocution, burns, fires, and death.
- Septic tanks and other confined space that may contain hazardous atmosphere, including insufficient oxygen, toxic gases, or particles.
- Wastewater stabilization ponds, aeration tanks or ponds, can pose a danger of drowning or falling.
- Other hazards that can cause injuries, such as from animal bites, stings, steps, dehydration, and excessive exposure to sunburn.

## PRACTICE SAFE WORK

Safe work practices help control hazards and manage risks associated with the provision and maintenance of scheduled desludging services. All service providers should comply with general safe work practice prevention principles[1] as follows:

1. Use anti-slip safety shoes with insulated sole.
2. Wear personal protecting equipment and clothing that can protect individuals from chemical emergencies from hazards affecting respiratory system, skin, eyes, hands, feet, head, body, and hearing.
3. DO NOT mix chemicals without the proper supervision of a qualified chemist or professional safety staff.
4. Follow all safety instructions regarding storage, transportation, handling or chemical pouring.
5. Check electrical equipment before use, verify that all power cords are properly insulated, replace damaged electrical equipment or ask an electrician for testing and repair.
6. Wear safety glasses in all cases where the eyes might be exposed to dust, flying particles, or splashes of dangerous liquids.
7. Wear respirators, or gas masks, when exposed to aerosol, dust, vapor or gas hazardous materials.
8. Be careful when handling highly corrosive liquid substances or chlorine, concentrated acids or alkalis, or when toxic gases can be emitted from reagents.
9. Comply with all safety instructions regarding entry to confined/isolated/special/narrow spaces, for example checking oxygen or toxic gas levels, using respiratory protective equipment if needed, there must be colleagues to guard if needed for assistance.
10. Do not smoke, eat or drink in areas where there may be a lot of chemical or biological contamination.
11. Use non-latex gloves if an allergy to latex material has been diagnosed.
12. All workers must undergo periodic examinations by the occupational physician to determine the initial symptoms of a possible chronic or allergic effect.
13. Learn to safely use levers and techniques to move heavy or hazardous loads such as chemical containers; use mechanical aids to help in lifting weights.
14. Maintain the completeness of the first-aid kit and provide training on how to use it.
15. Post telephone numbers for emergency response (police, fire department, ambulance) at important locations around the installation.

## MONITOR CRITICAL AREAS

There are critical areas that must be addressed for a comprehensive and effective safety approach in the provision and maintenance of scheduled desludging services. They include plans and action that help the safety engine run without glitches.

1. **Have the ability to assess risks and danger.** A danger is anything that can cause damage, such as exposure to septage, turn over corrosive chemicals such as acids or bases, and others. Estimating a hazard or risk, high or low, that a person can be harmed by a hazard, along with an indication of how serious the hazard is.

---

[1]ILO's Encyclopedia of Occupational Safety and Health, 4th Edition, Stellman, J. Mager, Editor, Vol. 2, pp. 55.29-32, 35–39 (1998). http://www.ilo.org/wcmsp5/groups/public/---ed_ protect /--- protrav /--- safework / documents / publication / wcms_192394.pdf

2. **Use of personal protecting equipment.** It promotes a sense of professionalism and a culture of safety. Workers should wear personal protecting equipment at any time they do their work. Not only during inspection activities.
3. **Promoting hygiene.** Septage contains infectious materials. Hygiene must be practiced at all steps in the collection, processing and disposal of septage. Avoid unnecessary contact with feces.
4. **Electricity security.** Septage treatment plants often have many electrical components, some using high voltage and electric current. Electric shock can be deadly. Only trained and certified workers are allowed to carry out repairs to electrical equipment.
5. **Entering a narrow-confined space.** A narrow-confined space includes a basement, tank, or inspection hole. If the confined space cannot be made safe for the worker by taking precautions, then workers should not enter the space until it is made safe to enter by additional means. All confined spaces should be considered hazardous unless a competent person has determined otherwise through a risk assessment. A confined space must also be controlled by a licensing program and should always use warning signs that are familiar and easily seen.
6. **Unintentional drowning.** To process septage, most septage treatment plants use ponds or lagoons. To prevent from unintentional drowning, safety equipment such as ring buoy, rope, hook, etc. should be provided.
7. **Vehicle safety.** Septage collection and transportation activities require special attention to safety. Dangers can occur due to vehicle maintenance, collection of sludge, driving a vehicle and transportation and traffic in and around treatment plants.
8. **Chemicals safety.** Chemicals are often used in septage treatment plants. Sometimes it uses a lot of chemicals, some of which pose a serious danger. Some of the most common chemicals used in septage treatment plants include chlorine (in one or more of its forms), sulfur dioxide, lime, polymer, methane and methanol, ferric chloride, alum, ammonia, acid (sulfuric, chloride, nitrate), bases (ammonium hydroxide, sodium hydroxide). They may also have a laboratory on site, with various chemicals used for process or quality testing.
9. **Public safety and security.** Septage treatment plants are often visited by community members, school classes, people from nearby cities, or foreigners from different countries. Visitors pose special risks to occupational health and safety because they are often not familiar with the dangers in the place of visit. Also, while observing desludging operations, including family group members (particularly children) should not be too close to hoses, trucks and septic tanks that are being emptied, to avoid direct contact with fresh sludge. Spills or leaks must be cleaned immediately and given a disinfectant.

IWA Publishing's authorised EU representative for General Product Safety Regulations is Diane D'Arras, 15 rue Duret, 75116 Paris, France, e-mail: safety@iwap.co.uk.

Printed and bound by CPI Group (UK) Ltd, Croydon, CR0 4YY
12/05/2026
02108869-0020